Cortical Circuits

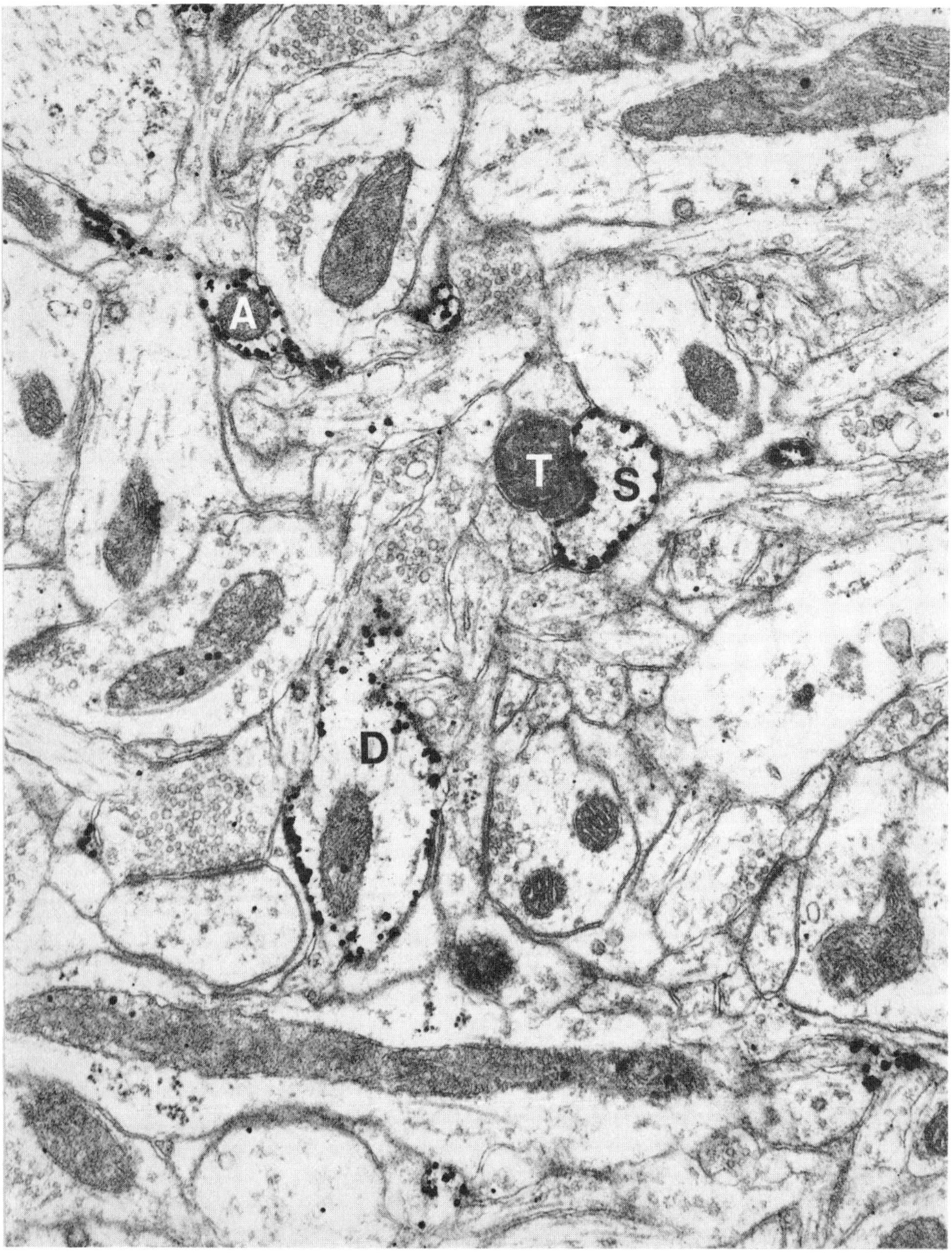

Electron micrograph of a thin section through the neuropil of layer IV in the mouse primary somatosensory cortex. Identifying the types of neurons, and the numbers and distribution of the different types of synapses that connect them with each other, is the essence of the study of synaptic organization. To obtain a hint of the complexity of this undertaking, the reader should consider that the volume of tissue depicted in this electron micrograph, which shows approximately 120 neuronal elements, represents only about 10^{-11} of the volume of the mouse cerebral cortex. Of the elements shown, four are labeled: a dendrite (D) belonging to a layer IV spiny stellate neuron, a spine (S) belonging to this dendrite, a degenerating thalamocortical axon terminal (T) that synapses onto the spine (S), and an axon (A) from an unknown source. Magnification × 30,000.

Edward L. White

Cortical Circuits

Synaptic Organization of the Cerebral Cortex Structure, Function, and Theory

With Asaf Keller

Introduction by Thomas A. Woolsey

With 31 Illustrations

Birkhäuser
Boston • Basel

Edward L. White, Ph.D.
Department of Morphology
Corob Center for Medical Sciences
Ben Gurion University of the Negev
Beer Sheva, Israel

Asaf Keller, Ph.D.
Department of Morphology
Corob Center for Medical Sciences
Ben Gurion University of the Negev
Beer Sheva, Israel

Current address
The Rockefeller University
1230 York Avenue
New York, NY 10021, USA

Library of Congress Cataloging-in-Publication Data
White, Edward L.
Cortical circuits: synaptic organization of the cerebral cortex—structure, function, and theory/Edward L. White; with contributions by Asaf Keller; introduction by Thomas A. Woolsey.
p. cm.
Includes bibliographies and index.
ISBN (invalid) 0-8176-03402-9 (alk. paper)
1. Cerebral cortex. 2. Neural circuitry. 3. Synapses.
I. Keller, Asaf. II. Title.
[DNLM: 1. Cerebral Cortex—anatomy & histology. 2. Cerebral Cortex—physiology. WL 307 W583c]
QP383.W45 1989
599'.0188—dc19
DNLM/DLC 88-37182

Printed on acid-free paper.

Typeset by David Seham Associates, Inc., Metuchen, New Jersey.
Printed and bound by Edwards Brothers, Inc., Ann Arbor, Michigan.
Printed in the United States of America.

9 8 7 6 5 4 3 2 1

ISBN 0-8176-03402-9 Birkhäuser Boston, Inc., Boston
ISBN 3-7643-03402-9 Birkhäuser Verlag, Basel

This book is dedicated to my adviser, Thomas S. Reese, M.D., and to a brave soldier, Paul Shane, Ph.D.

Acknowledgments

This book represents far more than the labor of its authors. In particular, we are indebted to Professor Thomas A. Woolsey for carefully reviewing the manuscript. His comments on content and style and his suggestions for the statistical treatment of our results have made this book eminently more readable and worthwhile than it would have been. I am grateful to Professor Moshe Abeles for suggesting that this book be written and for encouraging me to continue with the work when I simply didn't feel like it. Professor Abeles' interpretations of my data continue to be invaluable, and many of the novel ideas in this book are far more his than mine. We thank Professor Daniel J. Simons and Ms. Edith Elhanany for their valuable comments on the data and on the manuscript. My colleagues and I are particularly grateful to Sidney and Elizabeth Corob of London, England, for bringing us out of the desert by providing us with laboratories, and for their continued generosity toward us.

While writing these acknowledgments, I cannot help but think of certain teachers and colleagues who provided me with invaluable help along the way: For their inspiration, I thank Professors Vernon Mountcastle, Alan Peters, and Clinton N. Woolsey; for their instruction, counsel, and example, I thank my mentors, Professors Joseph C. Curtis, David G. Moulton, and Thomas S. Reese; for their illuminating comments, I am grateful to Professors Michael J. Gutnick, Michael T. Shipley, and Hendrik Van der Loos; for their support and many kindnesses, I thank Professors Milton W. Brightman and Dominic Purpura; and for their friendship and *menschlichkeit,* I thank Professors Reuven Yagil, Marc Colonnier, and Ben Cipolloni. These categories are not mutually exclusive.

I am most grateful to my wife, Roz, and to my children, Temira, Avner,

Avishai, and, recently, Amiram, for putting up with the long hours required to produce this book and for my moods engendered by it; to my parents, Saul and Hazel White, for their struggles to educate me; and to Leonard M. White and Robert E. White, for their generosity. Finally, I thank Miki Friedman of Tel Aviv for his help with names and titles.

Beer Sheva, Israel *Edward L. White, Ph.D.*

Contents

Introduction

This elegant book presents current evidence on the organization of the mammalian cerebral cortex. The focus on synapses and their function provides the basis for understanding how this critical part of the brain could work. Dr. White and his colleague Dr. Keller have collated an impressive mass of material. This makes the crucial information accessible and coherent.

Dr. White pioneered an area of investigation that to most others, and occasionally to himself, seemed a bottomless pit of painstaking attention to detail for the identification and enumeration of cortical synapses. I do not recall that he or anyone else suspected, when he began to publish his now classic papers, that the work would be central to an accelerating convergence of information and ideas from neurobiology and computer science, especially artificial intelligence (AI) (Rumelhart and McClelland, 1986).

The brain is the principal organ responsible for the adaptive capacities of animals. What has impressed students of biology, of medicine, and, to an extent, of philosophy is the correlation between the prominence of the cerebral cortex and the adaptive "complexity" of a particular species. Most agree that the cortex is what sets Homo sapiens apart from other species quantitatively and qualitatively (Rakic, 1988). This is summarized in the first chapter.

The phrenologists named different cortical regions, which they considered different organs of the mind (Spurzheim, 1824). This served as a stimulus for scientists to investigate regional variations of the cortex (Betz, 1874; Hughlings Jackson, 1958; Wilkins, 1963). The resultant concepts of localization of function (Ferrier, 1876; Grünbaum and Sherrington, 1903) and of the relation of function to cortical structure (Campbell, 1905) are cornerstones of neurological and physiological thought (Dandy, 1945).

Separate cortical areas, with proven different functions and anatomies, continue to occupy the attention of neuroscientists (Newsome and Paré, 1988). New findings from studies of the human brain are opening up long-sought vistas. Cortical areas that are involved in uniquely human activities (Petersen et al., 1988) and in apparently uniquely human problems (Reiman, 1987) are being discovered and will no doubt continue to be found. It has been argued that these areas and certain anatomical units that compose them (what defines these "units" is still debated) are unique and possibly genetically specified (Rakic, 1988). In the extreme, this evidence could mean that each area of the cortex represents the modern-day equivalent of the phrenologists' organs of the mind. If each is different, then each must be studied for its own sake, to know how each organ operates. This is a premise that can be addressed directly.

There are certain characteristics of the separate cortical areas that are similar if not identical. Many have a systematic representation or map of some functional attribute, of which motor and sensory maps are still the clearest examples (Woolsey, 1981). Most, if not all, have functionally demonstrated vertical arrays of neurons generally concerned with similar aspects of a particular function. Within such an array, each neuron has special characteristics related to its relative depth in the cortical mantle (Mountcastle, 1978). This and other evidence could be taken to mean that all cortex is fundamentally the same (as much as are neuromuscular junctions on different types of muscles). If all cortex is the same, then an understanding of a few areas in detail, just as understanding synapses at a few muscles in detail, ought to lead to general principles that can be invoked to explain all cortex. These views are likewise subject to scientific scrutiny. *Cortical Circuits* compares the evidence for special and general properties of the cortex.

Opinions on regional cortical uniqueness versus uniform general cortical organization can be considered as one axis of thought. Another perpendicular axis of thought is one that varies in its assessment of the importance of quantitation. It could be argued that detailed quantitation of any particular aspect of this part of the brain is irrelevant to understanding it. Or such detailed information could be regarded as essential to understanding the cortex. Up to the present, it has not been possible to extract from the cortex any information leading to general laws (Bok, 1936; Rall, 1962), in the sense that the mathematical description of a falling apple can be generalized to predict the behavior of other projectiles. *Cortical Circuits* highlights the value of quantitative information for defining features in the tissue from which general laws of cortical organization could be derived.

One approach clearly summarized in Part I of this book is that of

Sholl (1956). He argued that the connectivity of the cerebral cortex should be considered statistically. This limits the problem of anatomical complexity by eliminating, as impossible, any interactions that do not meet certain minimal distance criteria. Only elements in the cortex meeting such criteria are candidates for intracortical articulation for processing. Sholl's book extended the precocious qualitative outline of Lorente de Nó (1938) to a consideration of probable interactions that are a basis for cortical function. These probable connections constitute a baseline. It is supposed that each element on this baseline can be modified by the action of a synapse. In contrast to the largely fail-safe, all-or-none neuromuscular synapse, these intracortical synapses partake in stochastic processes in which a synapse is on average more or less effective in moving a neuron to or from threshold depending on the temporal and spatial sequences of, the sign of, and the temporal history of its inputs. Integration then occurs to smooth, shift, and grade performance by the output of variable groups of cortical neurons, which in some situations may be said to learn (Edelman, 1987). Without the hard data, these arguments are like the summer air over the Dead Sea.

This book summarizes a wealth of information relevant to the intrinsic organization of the cerebral neocortex. In Part I the chapters give an account of the anatomical characteristics of cortical areas, the principal cell types in the cortex, their synaptic relationships, and their known connectivity. Here, Dr. White reviews the data critically with functional interpretations reserved for later. The next three chapters, in Part II, are Dr. Keller's summary of aspects of the neurophysiological literature against the background set by Dr. White. The central feature is a critical review of the data with consideration of the structural bases for the functional findings. Part III brings together these methodologically defined separate streams. Both men working in concert generate a synthesis of the data. This synthesis has many aspects that extend beyond the immediate topic of the cerebral cortex and has far-reaching consequences for students of the brain, of biology, and, most certainly, of areas in computer science.

This book compiles carefully evaluated information that is difficult to glean from other larger works. Much of the information will be familiar to neuroscientists, although they will find its presentation new. For those outside neuroscience, the reading is facilitated through the use of figures and marginalia. Here in one volume are the facts basic to our current understanding of the mammalian cerebral cortex. *Cortical Circuits* ought to be read and studied by anyone who wants to know how the brain works.

St. Louis, Missouri *Thomas A. Woolsey, M.D.*

References

Betz, V. A. (1874). Anatomischer Nachweis zweier Gehrincentra. *Zbl. med. Wiss 12,* 578–580.

Bok, S. T. (1936). A quantitative analysis of the structure of the cerebral cortex. *Proc. Acad. Sci. Amst. 35,* 1–55.

Campbell, A. W. (1905). *Histological Studies on the Localization of Cerebral Function.* Cambridge University Press, Cambridge.

Dandy, W. E. (1945). *Surgery of the Brain.* Prior, Hagerstown, Md.

Edelman, G. M. (1987). *Neural Darwinism: The Theory of Neuronal Group Selection.* Basic Books, New York.

Ferrier, D. (1876). *The Functions of the Brain.* Smith, Elder & Co., London.

Grünbaum, A. S. F., and Sherrington, C. S. (1903). Observations on the physiology of the cerebral cortex of the anthropoid apes. *Proc. R. Soc. 72,* 152–155.

Hughlings Jackson, J. (1958). *Selected Writings of John Hughlings Jackson,* Vol. 1 (J. Taylor, ed.). Basic Books, New York.

Lorente de Nó, R. (1938). Architectonics and structure of the cerebral cortex. In: *Physiology of the Nervous System* (J. F. Fulton, ed.). Oxford University Press, New York, pp. 291–327.

Meynert, T. (1872). Vom Gehirn der Säugetiere. In: *Handbuch der Lehre von dem Geweben des Menschen und der Thiere* (S. Stricker, ed.). W. Engelman, Leipzig.

Mountcastle, V. B. (1978). An organizing principle for cerebral function: The unit module and the distributed system. In: *The Mindful Brain: Cortical Organization and the Group-Selective Theory of Higher Brain Function* (G. M. Edelman and V. B. Mountcastle, eds.). MIT Press, Cambridge, Mass., pp. 7–50.

Newsome, W. T., and Paré, E. B. (1988). A selective impairment of motion perception following lesions of the middle temporal visual area (MT). *J. Neurosci. 8,* 2201–2211.

Petersen, S. E., Fox, P. T., Posner, M. I., Mintun, M., and Raichle, M. E. (1988). Positron emission tomographic studies of the cortical anatomy of single word processing. *Nature 331,* 585–589.

Rakic, P. (1988). Specification of cerebral cortical areas. *Science 241,* 170–176.

Rall, W. (1962). Theory of physiological properties of dendrites. *Ann. N.Y. Acad. Sci. 96,* 1071–1092.

Reiman, E. M. (1987). The study of panic disorder using positron emission tomography. *Psychiat. Devel. 1,* 63–78.

Rumelhart, D. E., McClelland, J. L., and the PDP Research Group (1986). *Parallel Distributed Processing: Explorations in the Microstructure of Cognition,* 2 vols. MIT Press, Cambridge, Mass.

Sholl, D. A. (1956). *The Organization of the Cerebral Cortex.* Methuen, London.

Spurzheim, G. (1824). *The Anatomy of the Brain, With a General View of the Nervous System.* S. Highley, London.

Wilkins, H. (1963). Neurosurgical classics XII. *J. Neurosurg. 20,* 904–916.

Woolsey, C. N. (1981). *Cortical Sensory Organization,* 3 vols. Humana, Clifton, N.J.

Part I Anatomy

EDWARD L. WHITE

The function of any organ depends on interactions of its constituent cells with each other and with their environment; in the cerebral cortex, these interactions involve neuronal, glial, and vascular elements. Because the processing of information within the cortex is most closely associated with neuronal elements, this book will focus on neuronal interactions. Current knowledge suggests that communication between neurons takes place primarily at synaptic junctions. For this reason, an understanding of the function of the cerebral cortex hinges on the ability to identify its constituent neurons and to determine the sources, the numbers, and the types of synapses linking them with each other and with neurons in other parts of the brain.

The theoretical framework upon which this and subsequent chapters are based derives from the proposition that the processing of information by the nervous system is effected through synapses, and that the anatomical organization of synaptic connections, that is, the type of synapse and the identities of the neurons involved, is largely stable within the individual adult animal and relatively constant from one species to the next. In contrast, the physiology of synapses is plastic, allowing for variations on a scale of milliseconds in the level of electrical activity within synaptic pathways. Indeed, it can be assumed that the activity of synaptic pathways changes constantly, reflecting as it were the multitude of interactions between different neurons whose own levels of activity are in a constant state of flux.

The notion that the cerebral cortex of different species is built according to some common, basic plan is woven throughout the following treatise. Substantial evidence in support of this thesis has surfaced regularly since the outset of anatomical studies of cortical organization (e.g., see Cajal, 1909–1911, pp. 519, 824; Fleischhauer, 1978; Lorente de Nó, 1938) and has been confirmed by the many experimental approaches that have been used to elucidate different functional aspects of the cerebral cortex. For instance, direct electrical stimulation of the brain or recording of evoked potentials from it have established that the cerebral cortex can be parcellated into functional areas whose relative positions are similar in all mammalian species (e.g., Campos and Welker, 1976; Penfield and Rasmussen, 1950; and see reviews by Dykes and Ruest, 1986; Welker, 1976; Woolsey, 1964). Subsequently, recordings of single-unit activity have confirmed the existence of functional columns in various areas of the cortex in the mouse, rat, cat, monkey, and man (see Asanuma, 1975; Dykes, 1978; Hubel and Wiesel, 1977; Mountcastle, 1979) and have provided evidence that cortical neurons in a variety of species display similar functional properties (e.g., see Creutzfeldt, 1977, 1978).

The search for similarities in the anatomical organization of the cerebral cortex in different species began in earnest in the early 1900s (e.g., Cajal, 1909–1911) and was later pursued intensely by Lorente de Nó (1922, 1938). It is largely due to their efforts that we may now accept as valid the proposition that homologous functional areas of the cerebral cortex are organized into layers whose number, arrangement, and component morphological cell types and input and output pathways are strikingly similar in different species. More recent evidence for the similarity in the anatomical organization of the cerebral cortex in different species has been adduced by Rockel et al. (1980, and see Haug, 1987; Peters, 1987b): With the exception of area 17 in a number of primates, various regions of the cerebral cortex in mouse, rat, cat, monkey, and man are purported to contain essentially identical numbers of neurons beneath comparable areas of pial surface. In addition, the ratio of different neuronal types has been shown to be essentially the same in the motor and visual cortices of both lisencephalic and gyrencephalic brains (Winfield et al., 1980). The important relationship of synaptic circuitry to the hypothesis of the universality of cortical organization in different species is emphasized by Lorente de Nó's (1938) statement that "What remains constant [among different species and in different cortical areas] is the arrangement of the plexuses of dendritic and axonal branches, i.e., of the synaptic articulations through which nerve impulses are transmitted."

The recognition of neurons as discrete entities (see the review by Van der Loos, 1967) formed a basis for attempts to identify the pathways through which information passes from one nerve cell to another. The aim of Part I is to provide a comprehensive picture of the circuitry by which various neuronal types communicate. We shall begin with a brief, general picture of the organization of the cerebral cortex (Chapter 1) and of the morphology of the neuronal elements it contains (Chapter 2) and finish with a detailed presentation of the synaptic connections by which these neurons are linked (Chapter 3). This information will be used in subsequent chapters in an attempt to clarify various aspects of cortical function that are intimately related to the circuitry within this region. The principal focus of this work is the cerebral cortex because, perhaps more than any other area of the nervous system, it is the cerebral cortex that sets apart mammals from other animals and man from other mammals.

General Organization of the Cerebral Cortex 1

Cortical Areas

The cerebral cortex has been parcellated into a variety of regions according to differences in the local arrangements of cell bodies and myelinated fibers (e.g., Campbell, 1905; Rose, 1929; and see review by Kemper and Galaburda, 1984) and on the basis of studies that indicate that specific functions are localized to specific cortical areas (e.g., Penfield and Rasmussen, 1950; and see Kupfermann, 1985). Anatomical evidence for the parcellation of the cerebral cortex into different areas was adduced as early as the late 18th century by Gennari, who recognized a prominent white line running within the substance of the cerebral cortex parallel to the surface of the occipital lobe. Structures corresponding to this line or *stria* were observed in other regions of the cortex (Gennari, 1782), but nowhere was the line as clear as in the region that became known later as the striate cortex.

Subsequent anatomical studies established that the striate area is the terminus for a system of topographically organized projections that begin in the retina and proceed through one of the main sensory nuclei of the thalamus on their way to the occipital cortex (e.g., Minkowski, 1913; Polyak, 1933, 1957; Putnam and Putnam, 1926). Regions of the cortex that receive extensive input from the main sensory relay nuclei of the thalamus were termed *primary* or *first* projection areas, and accordingly, the striate cortex became recognized along with corresponding auditory and somatosensory regions as one of three primary projection areas (Rose and Woolsey, 1949, 1958).

Subsequent recordings in striate cortex of short latency or "primary" responses to visual stimuli led to the identification within this region of a *primary response* area that contained a topographic map of the visual field (e.g., Talbot and Marshall, 1941; Woolsey, 1947). Equivalent primary response areas were identified for the somatosensory (e.g., Walker, 1938; Woolsey, 1958; Woolsey and Fairman, 1946; Woolsey et al., 1942; Woolsey, 1967) and auditory (Ades, 1959; Rose and Woolsey, 1958; Walker, 1938; Woolsey and Walzl, 1942; Woolsey, 1967) systems. In time, the primary response areas as determined by studies of cortical evoked potentials came to be equated with the primary projection areas that had been defined by anatomical methods, so that today the term *primary* as applied to an area of the cerebral cortex implies primacy both in an anatomical and in a physiological sense.

The identification of the primary sensory areas (auditory, AI; somatosensory, SI; visual, VI—the striate area), and of a cytoarchitecturally distinct primary motor area (MI) where stimulation evokes discrete movements on the opposite side of the body (see, e.g., Woolsey, 1958, 1964), was soon followed by the discovery in a variety of species of additional representations of the periphery in other, "secondary" areas, whose cytoarchitecture differed from that of the primary areas (AII, SII, VII, MII; e.g., Woolsey, 1958, 1964). The use of the term *secondary* is somewhat unfortunate because of the implication that the secondary areas are in some way subordinate to the primary ones. The fact is that the secondary areas were so called simply because their discovery followed that of the primary areas (e.g., White, 1979, 1987).

Refinements in recording techniques in recent years have led to a concomitant explosion in the number of representations of the periphery that have been identified in the cerebral cortex. For the cat and monkey visual systems, this has meant the identification of nearly 20 separate cortical areas, outside of VI and VII, that are involved wholly, or in part, with the processing of visual information (see reviews by Rosenquist, 1985; Van Essen, 1985). Additional auditory areas also have been identified in regions of cortex outside of AI and AII (see the review by Brugge and Reale, 1985).

Similar observations have been made for the somatosensory and motor systems, but here the situation is somewhat different in that several separate representations of the periphery have been identified within the confines of regions previously identified as SI (e.g., Kaas et al., 1979; Merzenich et al., 1978; Sretavan and Dykes, 1983), SII (Robinson and Burton, 1980a, b), or MI (e.g., Pappas and Strick, 1981; Strick and Preston, 1978; Tanji and Wise, 1981). It is still far too early to determine if the *addition* of multiple areas of representation to areas VI and VII and to AI and AII, as compared with the *subdivision* of areas SI, SII, and MI into multiple areas of representation, reflects a basic difference in the organization of cortical regions devoted to the visual and auditory systems on the one hand and the somatosensory and motor systems on the other. Alternatively, it may be that the observation of different patterns of representation is related to the fact that, in each system, the representation of the periphery has been investigated using different experimental approaches.

The various sensory and motor areas of the cerebral cortex differ from one another in a number of ways, including the precise portion of the periphery that is represented, the topographic organization of the periphery, and the amount of cortical tissue that is devoted to this representation. Multiple representations in sensory and motor regions of the cerebral cortex may also differ in one or more of the following properties: cytoarchitecture, patterns of connectivity, physiological response properties, and behavioral effects of lesion or stimulation.

The manner in which multiple representations cooperate in the processing of cortical information is unclear. One possibility is that multiple representations are involved in hierarchical arrangements whereby input from the periphery is passed in stepwise fashion through a series of cortical areas, each one representing a more complex stage in cortical processing. In both sensory and motor systems, the different cortical areas are linked by reciprocal pathways; an asymmetry in the laminar organization of these connections in visual (Van Essen and Maunsell, 1983), somatosensory (e.g., Friedman, 1983), and auditory regions (Galaburda and Pandya, 1983) has been used to buttress arguments for the hierarchical processing of information by multiple cortical areas (see Van Essen, 1985, pp. 283–293).

Alternatively, the multiple representations may act in parallel to process separate aspects of input from the periphery (e.g., Ballard et al., 1983), a contention supported by results showing that different cortical areas receive input from different regions of the thalamus (e.g., for visual areas, Dreher et al., 1980; for somatosensory areas, Dykes, 1983). Evidence for hierarchical and parallel processing of information by the cortex will be dealt with in depth later, in Parts II and III.

Cortical association areas are considered to be highly integrative regions of cortex lying outside the primary and secondary sensory and motor areas (e.g., see Pandya and Yeterian, 1985). The relationship, and possible identity, of the association areas with some of the more recently identified areas of multiple representation is still unclear. The principal reason for this is that there is no general consensus as to what constitutes a cortical area or what criteria should be applied to identify and to distinguish between different cortical areas (see Van Essen, 1985). This deficiency is in part due to the nature of the different approaches used to study the cortex, but also to certain little-understood complexities in the organization of the brain. For instance, SI of the cat is composed of several different cytoarchitectonic subregions. Each subregion contains at least one representation of the periphery (see Dykes and Ruest, 1986), but one of the subregions contains two representations (Srevatan and Dykes, 1983) for which cytoarchitectural correlates have not been identified.

Summary

Topographically organized pathways project from sensory organs, through thalamic nuclei and on to cytoarchitecturally distinct areas of the cerebral cortex. The general correspondence of certain cytoarchitectonic areas (i.e., the primary and secondary sensory and motor areas) with physiological maps of the sensory or motor periphery supports the idea that the cerebral cortex is composed of cytoarchitecturally distinct zones involved in different functional activities (e.g., Walker, 1938). Correlations of structure and function in cortical areas are now being reassessed, however, principally because of the realization that single cytoarchitectonic areas may contain multiple representations of the periphery.

Lamination

Each cytoarchitectonic area can be subdivided into horizontal layers and occasionally into sublayers according to the types and packing density of the neuronal elements it contains. Elements contributing to patterns of cortical lamination include neuronal cell bodies, visualized traditionally in Nissl preparations, and axonal pathways, seen with various fiber stains, that originate within the same area of the cortex (i.e., intrinsic pathways) or that enter it from other cortical or subcortical regions (i.e., extrinsic pathways). Early efforts to describe a basic pattern of layering for the cerebral cortex disagreed both on the number of layers and on the boundaries between them, so that, by the beginning of this century, the basic number of cortical layers was variously argued to be five (e.g., Meynert, 1869–1872), six (e.g., Lewis, 1878), or seven (e.g., Cajal, 1909–1911, p. 521; Campbell, 1905); for the striate area, the number of layers reached eight (Meynert, 1869–1872) and even nine (Cajal, 1909–1911, p. 603). The early controversies regarding cortical lamination patterns stemmed, in part, from the fact that different workers examined sections from different parts of the brain using a variety of methods and approaches. Today, in line with the system employed by Brodmann (1906, 1909), the cerebral cortex in all mammalian species is divided into six basic layers (Figure 1.1).

The layers are named in rough accordance with the types of cell bodies they contain: I—plexiform or molecular layer; II—layer of small pyramids; III—layer of medium and large superficial pyramids; IV—the granular layer; V—layer of large deep pyramids; and VI—layer of polymorphic or fusiform cells, also termed the *multiform layer* (cf., Cajal, 1909–1911, p. 521; Lorente de Nó, 1938, p. 297). The basic arrangement of six layers is clearest in areas of the neocortex referred to by Brodmann (1909) as homotypic; the six-layered arrangement is somewhat modified in areas he labeled as heterotypic. Heterotypic cortex is divided into a granular type characteristic of sensory areas and an agranular type found in motor areas. In the former, the occurrence of many small cell bodies in layers II–IV makes it difficult to distinguish the borders between these layers; in the latter, the small cell bodies typical of layer IV are nearly absent.

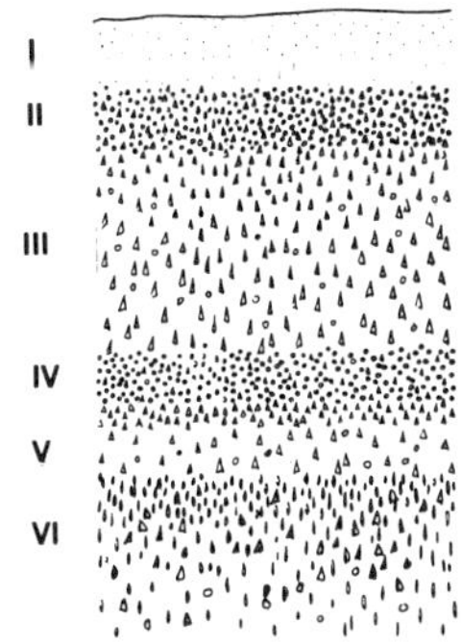

Figure 1.1. *Sketch providing a rough idea of how the six basic layers of the cerebral cortex are differentiated according to the relative numbers and packing density of pyramidal (triangular outlines) and nonpyramidal (round or ovoid outlines) neurons within them. (After Braak, 1984.) Reprinted with permission of Plenum Publishing Corporation from "Architectonics As Seen by Lipofuchsin Stains" by H. Braak, p. 86 in* Cerebral Cortex, *vol. 1, Peters, A., and Jones, E.G., eds. (1984).*

To account for local variations, the six basic laminae are often divided into sublayers. The ways in which laminae have been subdivided are well demonstrated by Table 1.1, which has been reprinted from the work of Braak (1984) to give the reader a flavor of the controversy surrounding efforts to define patterns of cortical lamination and to illustrate the different systems of labeling used in these efforts.

Also illustrated in Table 1.1 is the approximate laminar location of the stria of Gennari, which is now understood to be composed of myelinated axon collaterals belonging mainly to cell bodies situated in Brodmann's lamina IVb (e.g., Valverde, 1984). As outlined in the "Cortical Areas" section, the stria of Gennari is an identifying feature of the primary visual area of the cerebral cortex. This area was identified by Brodmann (1909) as a separate cytoarchitectural region, that is, area 17. This is just one example of many in which lamination patterns have been used as a major basis for the identification and differentiation of anatomically defined areas of the cerebral cortex.

The recognition of cortical laminae and the identification of their cellular components have had several consequences that are directly relevant to our consideration of cortical circuitry. For instance, the coexistence, and at times segregation, within a particular lamina of axons and dendrites belonging to specific neuronal types enables predictions to be made as to the kinds of synaptic connections that may be present within the lamina. As we shall see in Chapter 3, sometimes these predictions have not gone far enough or have been proved incorrect. Nevertheless, in many instances predictions of synaptic relationships based on an understanding of components of cortical laminae have provided useful insight into the functional organization of the cerebral cortex. For example, the main relay or "specific" nuclei of the thalamus that lie directly along the ascending sensory pathways from the periphery to

Table 1.1. *Pattern of lamination in the striate area of man and various subhuman primates according to various authors.*

Ramón y Cajal (1909–1911)	Brodmann (1909)	von Economo and Koskinas (1925)	Rose (1935)	Bailey and von Bonin (1951)	Garey (1971)	Lund and Boothe (1975)	Braak (1976)
1	I	I	I	i	I	I	I
2	II	II	II	ii	II	II	II
	III	III	III	iiia	IIIa	IIIA IIIB	IIIa,b
3	IVa	(IIIb)	IVa	iiib	IIIb	IVA	IIIc–IVa
4	IVb	(IIIc)	IVbα	iva	IIIc	IVB	IVb
			IVbβ			IVCα	IVcα
5	IVc	IV	IVc	ivbα	IV	IVCβ	IVcβ
6		Va	Va	ivbβ	V	VA	IVd
							Va
7	V	Vb	Vb	v		VB	Vb
8	VIa	VI	VI	via	VI	VI	VIa
9	VIb		VII	vib			VIb

[a]The dotted area gives the approximate location of the line of Gennari.

Reprinted with permission of Plenum Publishing Corporation from "Architectonics As Seen by Lipofuchsin Stains" by H. Braak, p. 72 in *Cerebral Cortex*, vol. 1, Peters, A., and Jones, E.G., eds. (1984).

the cortex have projections that terminate mainly within layer IV of the primary sensory areas (cf. Lorente de Nó, 1938). Layer IV also contains neurons whose cell bodies and dendrites are almost entirely restricted to this layer, and for this reason it was postulated that the layer IV cells receive large numbers of synapses from the specific thalamic projections (e.g., Lorente de Nó, 1938, p. 305). The value of this prediction is illustrated by findings showing that the response properties of neurons in layer IV are frequently similar to those of neurons within the specific thalamic nuclei (e.g., Chapin, 1986; Gilbert, 1977; Hubel and Wiesel, 1968, 1977; Simons, 1978, 1985).

Another example in which evaluations of laminar arrangements have been used for their predictive value has been in the development of hierarchical schemes between different cortical areas. Corticocortical projections from the primary visual area of the monkey originate nearly exclusively from neurons having cell bodies in the superficial layers of the cortex and terminate mainly within layer IV of the target region (see the review by Van Essen, 1985). By virtue of the similarity of this arrangement with that of thalamocortical projections, which also terminate preferentially in layer IV, the projections from the primary visual area to other visual areas were considered feedforward or ascending. Projections from the target regions back to the primary area tend to originate from deep, as well as from superficial, layers of the cortex, and to terminate, to a large extent, within laminae outside layer IV. These return projections were considered to be feedback or descending. Using similar considerations, ascending and descending polarities were assigned to the reciprocal connections that link 12 areas of visual representation in the monkey (Maunsell and Van Essen, 1983; Van Essen and Maunsell, 1983). Eventually, this hierarchical arrangement was extended to include 17 different areas of visual representation (Van Essen, 1985). In like fashion, the specifics of interlaminar connections have been used to postulate a hierarchical arrangement between different laminae within the primary visual cortex (e.g., Van Essen, 1985). In addition, the hierarchical ordering of neurons in both primary visual and somatosensory areas has been proposed on the basis of laminar differences in their receptive field properties (e.g., Chapin, 1986; Gilbert, 1977; Hubel and Wiesel, 1962, 1968; Simons, 1978; see also Part III).

Summary

The identification of cortical laminae as distinct entities stems in large part from the concentration within them of particular types of neuronal cell bodies and processes that arise locally or that enter the area from other regions of the brain. In general, the cerebral cortex is divided into six basic layers; typically, classification systems for cortical neurons tend to include some reference to the lamina containing the parent cell body as part of the appellation of the neuron (i.e., layer III pyramid).

Studies of the local variations in the appearance of cortical laminae have played a crucial role in the division of the cerebral cortex into different areas, a process that has contributed significantly to our understanding of the cerebral cortex as an organ in which specific functions are localized to specific regions. Conclusions based on the composition and patterns of connectivity of cortical laminae have great predictive value for synaptic relationships and show promise for interpreting functional relationships between different cortical areas and between different laminae within a single cortical area.

Vertical Arrangements

Coexisting with the horizontal, laminar aspect of cortical organization is a vertical one envisioned by Lorente de Nó (1938) to consist of chains of interconnected neurons extending across all the layers of the cortex. Confirmation of the notion that the cerebral cortex has a distinct vertical aspect to its organization was provided subsequently by the observation of functional columns in both sensory and motor areas (e.g., Asanuma, 1975; Dykes, 1978; Hubel and Wiesel, 1977; Mountcastle, 1957, 1979). Functional columns, demonstrated initially by electrophysiological methods, are composed, in sensory areas, of neurons that share similar response properties. For instance, there are vertical arrangements of cells in visual cortex that respond preferentially to stimuli presented to one eye or to stimuli having a particular orientation (Hubel and Wiesel, 1977). Vertical arrays of units in auditory cortex tend to respond to the same or to similar frequencies (e.g., Abeles and Goldstein, 1970), and vertical columns of neurons in somatosensory cortex respond to nearly identical peripheral receptive fields (e.g., Mountcastle, 1957; 1979). Columnar arrangements also have been identified in motor cortex where stimulation of discrete groups of neurons causes specific muscles to contract (Asanuma, 1975; Asanuma and Rosén, 1972).

Attempts to identify anatomical correlates of functional columns have employed a variety of approaches including the labeling of neuronal elements by anterograde degeneration (e.g., Hubel and Wiesel, 1972; Jones et al., 1979; Kosar and Hand, 1981), by the anterograde axonal transport of labeled proteins applied to the sensory periphery (e.g., Wiesel et al., 1974), to the thalamus (e.g., Curcio and Harting, 1978), or to areas of the cerebral cortex itself (e.g., Goldman and Nauta, 1977; Goldman-Rakic and Schwartz, 1982), by the intracortical injection of horseradish peroxidase (e.g., Rockland and Lund, 1983), or by the demonstration of patterns of metabolic activity as evidenced by the distribution of cytochrome oxidase (e.g., Land and Simons, 1985; Tootell et al., 1982; Wong-Riley and Welt, 1980) or radioactively tagged 2-deoxyglucose (e.g., Hand et al., 1979; Hubel et al., 1978; Kennedy et al., 1976; Lowel et al., 1987).

These efforts have resulted in the following discoveries: (1) Functional columns in many areas of the cerebral cortex may be shaped more like bands or slabs than like columns; (2) columnar arrangements based on the distribution of afferent fibers are a general feature of cortical organization; and (3) in most instances, clear structural correlates for functional columns have yet to be identified.

The correlation between functional columns and anatomy is most evident in the primary somatosensory cortex of mice and rats where multicellular columnar structures, termed barrels, can be visualized without the use of experimental methods (e.g., Woolsey and Van der Loos, 1970). Barrels, which occur only within layer IV, are coincident with functional columns that extend throughout the full thickness of the cortex (e.g., Simons, 1978; Simons and Woolsey, 1979). Despite the clarity and ease with which columnar structure (i.e., the barrels) can be identified within layer IV of rodent somatosensory cortex, parts of functional columns in layers of cortex above and below layer IV can be visualized only by experimental methods (Chmielowska and Kossut, 1984; Durham and Woolsey, 1977, 1978, 1985; Gonzalez and Sharp, 1985).

Summary

All areas of the cerebral cortex are characterized by certain prominent, vertically oriented elements that no doubt play a significant role in the formation of functional columns. Conspicuous among these elements are long, vertically oriented dendrites that are sometimes grouped into bundles (see, e.g., Peters and Kara, 1987; Peters and Walsh, 1972), and vertically oriented afferent, efferent, and intracortical axonal projections. Both dendrites and axons typically cut across several laminar boundaries in their passage through the cortex.

Clear structural correlates for functional columns are difficult to find. One possibility for the general lack of correspondence between anatomical structure and functional columns probably has to do with the likelihood that certain functional columns are rather ephemeral, owing their existence to the continued presence of a specific set of stimulus conditions (cf. Crawford, 1985; see also Part III, Chapter 7.)

Neurotransmitters and Receptors in the Cerebral Cortex

Any description of cortical circuitry and of the intercellular interactions within circuits would be incomplete without an identification of the neurotransmitters utilized by the neurons within these circuits and an understanding of the nature of the synaptic interactions mediated by these transmitters. In the last decade, several neurotransmitters, putative transmitters, and neuropeptides have been identified in the cerebral cortex (see reviews by Emson and Hunt, 1984; Krnjević, 1984; Taylor and Stone, 1981). Studies of the synaptic interactions involving neurons utilizing these neuroactive substances are few, and these have focused mainly on neurons containing gamma aminobutyric acid (GABA), which is considered to act as an inhibitory neurotransmitter (see review by Houser et al., 1984). To date, GABA has been localized only with axon terminals that form symmetrical synapses or within cell bodies and dendrites of neurons whose axons form symmetrical synapses (see Figure 1.2). Glutamate and aspartate, which are considered to be excitatory neurotransmitters, are associated only with types of neurons whose axons form asymmetrical synapses (see review by Streit, 1984). The association of GABA with symmetrical synapses and of glutamate and aspartate with asymmetrical synapses is consistent with the supposition that, in general, symmetrical synapses are inhibitory, whereas asymmetrical synapses are excitatory (see, e.g., Peters et al., 1976, pp. 148–150).

Glutamate and aspartate have been implicated as the excitatory neurotransmitters used by thalamocortical afferents to the primary sensory areas of the cerebral cortex (e.g., Ottersen et al., 1983; Tsumoto et al., 1986). Other neurotransmitters such as acetylcholine and the monoamines have been shown to participate in widespread afferent pathways

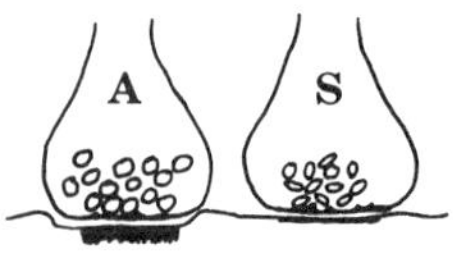

Figure 1.2. *The terms* symmetrical *and* asymmetrical *were applied by Colonnier (1968) to describe the two most frequent morphological types of chemical synapse observed in aldehyde-fixed cerebral cortex. Both types consist of a clustering of vesicles at the presynaptic membrane, a synaptic cleft, and a postsynaptic membrane.*

Asymmetrical synapses, roughly the equivalent of Gray's type I (Gray, 1959), have spherical vesicles approximately 50 nm in diameter and synaptic clefts approximately 30 nm wide. The most distinguishing feature of an asymmetrical synapse is a 40 to 50 nm thick band of electron dense material adherent to the cytoplasmic surface of the postsynaptic membrane. The absence of a similar band attached to the presynaptic membrane imparts an asymmetry to these synapses—hence the name asymmetrical.

Symmetrical synapses, roughly the equivalent of Gray's type II (Gray, 1959), have slightly smaller vesicles, which may appear oval shaped in thin sections, synaptic clefts that are approximately 20 nm wide, and a very thin plate of postsynaptic dense material (approximately 10 nm). The pre- and postsynaptic membranes, with their attached electron dense material, appear to be of roughly similar thickness, that is, symmetrical.

that innervate multiple cortical areas. These transmitters are believed to be involved in regulating the general level of synaptic activity (e.g., Bloom, 1981; Foote and Morrison, 1987), metabolism, and blood flow (e.g., Sillito and Murphy, 1987; Fallon and Loughlin, 1987) within the cerebral cortex. A discussion of the possible role in cortical function of acetylcholine and the monoamines will form the remainder of this section; information on other neuroactive substances, whose relationships to specific cortical circuits are more clearly defined, is presented in the sections of Chapter 2 that deal with specific neuronal types.

Axons containing the monoamines norepinephrine, dopamine, and serotonin have been shown to provide a dense innervation to a large number of cortical areas (e.g., McGeer et al., 1987; Morrison et al., 1984). For example, noradrenergic fibers innervate somatosensory, visual, motor, and temporal regions of the cerebral cortex (e.g., Morrison et al., 1984). Noradrenergic innervation is thought to modulate the activity of cortical circuitry by enhancing the selectivity and the responsiveness of neurons to specific sensory stimuli (Foote and Morrison, 1987; Waterhouse and Woodward, 1980), an effect that may explain the involvement of noradrenergic innervation in the behavioral changes associated with the sleep–wake cycle (Aston-Jones et al., 1984). In addition, norepinephrine has been implicated in modulating synaptic plasticity in the visual (e.g., Gordon et al., 1988) and in the somatosensory (Levin et al., 1988) areas of the cerebral cortex.

Dopaminergic fibers occur mainly in the limbic and association areas of the cortex (see Lindvall and Bjorklund, 1984), but some sensory areas also receive them (e.g., Phillipson et al., 1987). The functional role of dopaminergic afferents to the cerebral cortex is obscure, but it has been hypothesized that they too modulate the state of responsiveness of cortical neurons (see Foote and Morrison, 1987) and that they are involved with higher integrative cortical functions (Fallon and Loughlin, 1987). Serotonergic fibers densely innervate the entire neocortical mantle (Lidov et al., 1980). Serotonergic innervation is believed to play a role similiar to that proposed for the noradrenergic fibers, namely, the modulation of the responsiveness of cortical neurons to sensory input, including the control of the sleep–wake cycle (see Foote and Morrison, 1987; McGeer et al., 1987).

An additional neurotransmitter implicated in modulating the responsiveness of cortical neurons is acetylcholine, which has been identified in afferents innervating a variety of cortical areas (e.g., Bear et al., 1985; McGeer et al., 1987). Acetylcholine has been shown to modify visually evoked responses in cortical neurons, without significantly reducing the response selectivity of the cells, suggesting that this transmitter modulates the signal-to-noise ratio of cortical activity and mediates some of the effects of arousal (Sillito and Kemp, 1983; Sillito and Murphy, 1987; Sato et al., 1987). In addition, acetylcholine has been implicated in the facilitation of experience-induced alterations in the response properties of cortical neurons, a process that may indicate a possible role for acetylcholine in learning and memory (Krnjević et al., 1971; Krnjević, 1984; Bear and Singer, 1986; Sillito and Murphy, 1987).

The identification of neuroactive substances within neurons provides useful information on the effects of presynaptic elements on their postsynaptic targets. Further insight into cortical circuitry can be attained by coupling this information with data on the location of the sites of neurotransmitter receptors. The localization of receptor sites should reveal neuronal populations that are potentially postsynaptic to their associated neurotransmitter. Receptors sites can be visualized by applying to cortical tissue a radiolabeled ligand that binds specifically to a particular receptor (see the review by Wamsley, 1984). However, the identification of a receptor site does not necessarily imply the juxtaposition of an axon terminal containing the neurotransmitter related to that receptor. This raises the possibility that certain neuroactive substances may be involved in extrasynaptic activation, reaching their receptor sites by diffusion through the cortical neuropil.

Each neurotransmitter must have at least one specific receptor, but many neurotransmitters are associated with multiple receptor types (see McGeer et al., 1987, Table 5.3). For example, it has been established that GABA operates via two distinct receptor subtypes, $GABA_A$ and $GABA_B$. The two GABA receptors differ both structurally and functionally, and they are sensitive to different agonists and antagonists (see the review by Bormann, 1988).

Summary

Monoamine and acetylcholine containing fibers are widely distributed to many cortical areas, where they are assumed to modulate the level of cortical activity. In general, the action of these neuromodulatory pathways is thought to provide a kind of "nonspecific" innervation that controls the overall responsiveness of broad areas of the cerebral cortex to "specific" sensory inputs. This contrast of "nonspecific" with "specific" innervation is reminiscent of an analogous situation described by Lorente de Nó (1938). He labeled as specific the projections from the main thalamic relay nuclei that terminate mainly within layer IV of the primary sensory areas. Projections from other thalamic nuclei were designated nonspecific because these projections terminate in all layers of more extensive regions of the cerebral cortex (see the "Lamination" section

Table 1.2. *Neuroactive substances identified in the cerebral cortex.*

Neurosubstance	Function	Anatomical distribution	References
Amino acids			
GABA	Inhibitory transmitter	Most classes of cortical nonpyramidal neurons	Houser et al., 1984; Jones et al., 1987
Glutamate and aspartate	Putative excitatory transmitters	Pyramidal neurons thalamocortical afferents	Ottersen et al., 1983; Streit, 1984; Tsumoto et al., 1986
Glycine	Putative inhibitory transmitter	Cortical projection to hypothalamus (?)	Kita and Oomura, 1982
Amines			
Acetylcholine	Modulation and regulation of cortical neuron activity	Fibers afferent to the cortex, a few cortical neurons	Bear et al., 1985; Parnavelas et al., 1986; Foote and Morrison, 1987; Sillito and Murphy, 1987
Histamine	?	Afferent pathways to the cortex	Garbag et al., 1980
Monoamines			
Norepinephrine, dopamine, and serotonin	Modulation and regulation of cortical neuron activity	Afferent pathways to the cortex	Foote and Morrison, 1987; Fallon and Loughlin, 1987
Peptides			
CCK, NPY, VIP, SP, and somatostatin	Modulation of synaptic transmission (?)	Cortical neurons, mostly nonpyramidal, most colocalize with GABA	Peters et al., 1987c; Emson and Hunt, 1984; Jones et al., 1987
Opioid peptides	Involved in nociceptive and analgesic mechanisms	Cell bodies and fibers in the cortex	Wamsley, 1984; Khachaturian et al., 1985

above). At present, it is difficult to assign precise functional roles to either type of "nonspecific" input mentioned above; clarification of these roles should be facilitated by determining the specific interactions between these afferents and identified cortical circuits.

Current data on the distribution and possible functions of neuroactive substances within the cerebral cortex are illustrated in Table 1.2. Certain of these substances will be mentioned again in Chapter 2 in relation to the pharmacology and function of specific neuronal types. For most of the neuroactive substances listed in Table 1.2, at least one receptor type has been identified in the cerebral cortex. For a comprehensive description of receptor localization studies in the cerebral cortex, the reader is referred to the reviews by Wamsley (1984) and McGeer et al. (1987).

Summary of Chapter 1

Cortical areas, patterns of lamination, vertical arrangements, and circuitry are all closely interrelated. In recent years it has become increasingly clear that every lamina, within each area of the cerebral cortex, is characterized by a unique combination of inputs, outputs, and intrinsic connections that project both horizontally and vertically across the laminae. Added to this are recent findings regarding the laminar specificity of neuroactive chemicals and their receptors. The specific complement of neuronal elements within a cortical layer determines the kinds of connections, that is, the circuits, that may occur within that layer. Knowledge of intra- and interlaminar circuitry is crucial for our understanding of cortical function. These synaptic interactions will be discussed later on in Chapter 3, but next it seems useful to describe, in some detail, the neuronal components involved in these circuits.

Cell Types 2

Cortical neurons are classified into two broad morphological categories: pyramidal, and stellate or nonpyramidal (e.g., Peters and Jones, 1984).

Pyramidal Cells

Pyramidal cells include a variety of morphological types that have the following features in common: A single, dominant apical dendrite, larger in diameter than the other dendrites, which usually extends vertically from the cell body toward the pial surface, and several basal dendrites, which radiate more or less horizontally from the base of the cell body (Figure 2.1). All the dendrites of pyramidal cells bear spines that tend to occur with greatest frequency in the middle regions of the dendrite (Globus and Scheibel, 1967). For a review of the possible role of spines in synaptic transmission and plasticity, see Coss and Perkel (1985) and Rall and Segev (1987). The axon of a pyramidal cell originates typically from the base of the cell body, or less frequently from the proximal portion of a basal dendrite, and projects into the white matter, giving off collateral branches on the way. Pyramidal cells with axons that do not enter the white matter have been observed in the cat, monkey, and human; in these instances the output of the pyramidal cell is presumed to be entirely local or to adjacent cortical areas (see Feldman, 1984; p. 161; Katz, 1987).

As Feldman points out in his comprehensive and careful review of pyramidal cell morphology (1984), the somata of pyramidal neurons exhibit a broad spectrum of shapes including spherical, ovoid, and

Figure 2.1. *Drawing of a Golgi-impregnated layer V pyramidal neuron from the primary somatosensory cortex of the mouse.*

rhomboidal in addition to the more classical conical shape. This is especially true of pyramidal neurons in layer VI where the variety of pyramidal cell shapes no doubt contributed to the naming of this layer as polymorphic or multiform. In fact, neither perikaryal size nor shape is reliable as the sole criterion for differentiating pyramidal from nonpyramidal cells (Feldman and Peters, 1978). Nevertheless, the application of the term *pyramidal cell* to the broad class of neurons sharing the features described above is generally accepted and has been confirmed by broad usage. For an account of pyramidal neurons that have been distinguished by morphological features including size (e.g., Betz cells), orientation (e.g., inverted pyramids), and laminar location (e.g., Meynert cells), the reader is referred to Feldman's review (1984, pp. 176–182).

A distinguishing aspect of pyramidal cell bodies is that they receive only symmetrical synapses; of the nonpyramidal cell types, only the spiny stellate neurons share this feature (e.g., Peters and Kaiserman-Abramof, 1970; White and Rock, 1980). The cell bodies of all other neuronal types receive both asymmetrical and symmetrical synapses (e.g., Peters and St. Marie, 1984, p. 435). The distribution of synaptic types onto dendrites is also similar for pyramidal and spiny stellate cells: Most synapses onto the dendritic shafts (axodendritic) of these neurons are symmetrical, whereas synapses onto their spines (axospinous) are usually of the asymmetrical type (e.g., Hersch and White, 1981b; White and Rock, 1980). In general, asymmetrical synapses form a comparatively large proportion of the axodendritic synapses onto other neuronal types (e.g., White et al., 1984).

Dendrites and Their Relationships With Cortical Laminae

From the perspective of intracortical circuitry, a particularly important aspect of the morphology of pyramidal neurons relates to the precise spatial relationships between specific cortical laminae and the cell bodies and processes of individual cortical pyramids. The importance of identifying lamina–neuronal interrelationships stems from the predictive value of these relationships for the existence of specific cortical circuits. The reason for this is simply that the formation of a synaptic connection requires the close apposition of pre- and postsynaptic elements, certainly the presence within a single lamina of both elements is required, and if the elements do not appose one another, a synapse cannot be formed. For instance, the apical dendrites of deep pyramidal cells in the primary sensory areas of the cortex frequently ascend through layer IV, where they have the opportunity to receive input from specific thalamic afferents that ramify within this layer; the dendrites of pyramidal neurons of layer II generally do not descend below layer III, and so these dendrites cannot receive synapses from thalamic afferents to layer IV.

Another instance in which pyramidal cell processes appear to be specifically distributed such that they preferentially receive some inputs but not others is exemplified by layer V pyramidal cells in area 17 of the monkey visual cortex. Both the apical and basal dendrites of these neurons branch extensively within layer V, and the apical dendrites emitted by the largest cell bodies also give off many oblique side branches in layer III. However, the apical dendrites display a conspicuous lack of side branches on their ascent through two sublaminae of layer IV, IVCβ, and IVA, and the spine density on the dendrites in sublayer IVCβ is unusually low relative to most other dendritic branches of these cells. Interestingly, layers IVA and IVCβ are the very sublayers that receive input from the parvocellular layers of the dorsal lateral geniculate body (Lund, 1973; Lund et al., 1979; Lund and Boothe 1975).

These findings do not exclude the possibility of a synaptic connection between this thalamocortical pathway and the layer V pyramids because their apical dendrites do pass through the sublayers in question, even if they do not arborize within them. In fact, as we shall see later on, an afferent pathway tends to form synapses with every neuronal element within its terminal field that is capable of receiving synapses of the appropriate morphological type. Still, at least intuitively, it would seem that pyramidal neurons that have apical dendrites that emit few branches within a specific layer are provided with less of an opportunity for synaptic interaction with laminar-specific axonal terminations than are cells whose apical dendrites branch profusely within the same layer.

Spatial relationships between pyramidal neurons and cortical laminae were investigated systematically by Lorente de Nó (1938), who observed that both the length of the apical dendrite and the specific cortical laminae within which its side branches occur can be correlated with the laminar location of the parent cell body. Lund and Boothe (1975) reexamined the question of lamina–neuronal relationships in great detail and showed that pyramidal neurons having somata within a single layer could be divided into subclasses according to the laminar distribution of side branches belonging to their apical dendrites. Subsequent results indicate that neurons within a morphological subclass of this type may project to a single target area. For example, Katz (1987) observed that in cat visual cortex apical dendrites of layer VI pyramids that project to the claustrum give off side branches mainly in layer V before terminating in layer I, whereas apical dendrites of layer VI pyramids that project to the lateral geniculate nucleus branch extensively in layers V and IV and terminate within or beneath the lower regions of layer III.

Specific lamina relationships also characterize the spatial distribution of basal dendrites belonging to pyramidal cells. For example, in all cortical layers, the basal dendrites of pyramidal cells tend to arborize horizontally and thus preferentially within the lamina containing their parent cell body. When this is not the case, the specificity of basal dendrite–lamina relationships is even more striking. For example, pyramidal cells of sublayer IVA in area 17 of the monkey cortex have basal dendrites that turn sharply downward from the cell body to enter sublayer IVB; they fan out horizontally in this layer, markedly avoiding any ramification within layer IVA (Lund, 1973; Lund and Boothe, 1975).

Collateral Axonal Branches

A general feature of pyramidal cells is that each neuron has, in addition to its long, projecting axon, a number of local axon collaterals that may leave the main trunk at any point along its course and travel horizontally or obliquely within the cortical laminae. Side branches of these axon collaterals may turn upward or downward to ramify within the cortex, at times in locales quite distant from the location of the parent cell body. For example, Gilbert and Wiesel (1979) describe a layer V pyramidal cell in cat visual cortex that has a local axon collateral that travels horizontally some 7 mm within layer VI. Long, intracortical collaterals were subsequently observed to be emitted by pyramidal cells in the supragranular layers of the visual cortex of the cat (Gilbert and Wiesel, 1981) and tree shrew (Rockland et al., 1982). In all of these instances, the collaterals remain within the cytoarchitectonic area of the parent cell body, but this need not be the case. For instance, DeFelipe et al. (1986a) have illustrated pyramidal cell axon collaterals that travel horizontally for relatively long distances before giving off side branches within an adjacent cytoarchitectonic area.

Before proceeding with a discussion of certain specifics regarding the collateralization of pyramidal cell axons, it would seem useful to provide a short description of the various kinds of axonal branches to which the word *collateral* has been applied. The term *collateral* originally was applied to those branches of an axon that ramify within the general vicinity of the cell body, the main axon being that which projects out of the immediate region of the cortex to other cortical areas or to subcortical regions of the brain (cf. the use of the terms *collatérale, collatérale récurrente,* and *cylindre-axe* by Cajal, 1909–1911). In the past few years, evidence has begun to accumulate that some pyramidal cell axons bifurcate and project to widely separate regions of the brain (e.g., Goldman-Rakic and Schwartz, 1982; Kennedy and Bullier, 1985). These cells might be considered to possess two main axons (and in this respect to resemble the nonpyramidal cells with double axons described by Meyer [1982] in cat visual cortex), but in fact, both axonal branches have been termed *collaterals.* Presumably in such instances, the main branch of the axon is that which occurs between the pyramidal cell body and the point at which the axon bifurcates.

A somewhat more serious conceptual difficulty regarding the identification of an axonal branch as collateral stems from the implication by use that a collateral branch is in some way subordinate to the main branch of the same axon. It is true that, in general, main axonal branches have larger diameters and travel for longer distances than their collaterals. However, it would seem logical that local axon collaterals are intimately involved in certain aspects of cortical processing, such as the intracortical shaping of receptive fields, and therefore, their importance for cortical processing should not be underestimated.

Until such time as the workings of the brain are more fully understood, when describing an axonal branch as a collateral it would be best to keep in mind the primary definitions of the word as given by *Webster's:* "side by side, in parallel." To simplify the discussion, we shall continue to refer to branches of a pyramidal cell axon that ramify within the immediate vicinity of the cell body as *local axon collaterals;* branches that project out of the immediate area of the cortex will be referred to as *projection axons.*

Early studies of the specific distribution of pyramidal cell axon collaterals relied exclusively on analyses of Golgi impregnated neurons, although it has long been claimed that a severe drawback of the Golgi method is the difficulty with which local axonal projections are labeled (e.g., Feldman, 1984, p. 161). Recent years have seen the increasing application of methods that employ the retrograde transport or intracellular injection of marker substances to label neurons, in part, because these approaches usually produce a more complete picture of the intrinsic collateralization patterns of the labeled neurons. Results obtained by means of the recently developed methods have confirmed and expanded upon previous results, but in no instance have findings obtained with the Golgi method been disproved. For this reason, in the following discussion of the collateralization patterns of pyramidal cell axons, no distinction has been made between results obtained by means of the different approaches to cell labeling.

The local axon collaterals of pyramidal cells exhibit a wide variety of branching patterns, and at first glance little order is discernible within them. However, systematic study of the ramifications of local axon collaterals lends considerable support to the belief that the distribution of axons within the cortex is orderly (e.g., Lorente de Nó, 1938; Lund and Boothe, 1975) and proceeds according to certain rules. For instance, pyramidal neurons typically have local axon collaterals that ramify extensively within close proximity to the parent cell body (e.g., in the primary visual cortex of the rat [Hedlich and Winkelmann, 1982; Parnavelas et al., 1977, 1983; Peters and Kara, 1985], cat [Gilbert and Wiesel, 1979; O'Leary, 1941], and monkey [Lund and Boothe, 1975]; in the secondary visual cortex of the monkey [Lund et al., 1981; Valverde, 1978]; in the primary auditory cortex of the cat [Winer, 1984]; and in the primary somatosensory cortex of the mouse [Lorente de Nó, 1922]). In addition to this usual arrangement, some pyramidal cells have collateral axonal branches that ascend, descend, or travel for long horizontal distances within the cortex.

The occurrence of these arrangements appears to be related in some consistent way to the laminar location of the parent cell body or to the terminal site to which its principal axon projects. For example, pyramidal cells having somata in layers V and VI often have ascending collaterals that ramify extensively within layers II and III (e.g., Cajal, 1909–1911; Lorente de Nó, 1938). Conversely, pyramidal cells of layers II and III typically send extensive local collateral projections to layer V (Figure 2.2), an arrangement that has been reported for the primary visual cortex in the rat (Parnavelas et al., 1977), cat (e.g., Gilbert and Wiesel, 1979, 1981; Kisvarday et al., 1986c), and monkey (Tigges and Tigges, 1982; Lund and Boothe, 1975); for the secondary visual cortex in the monkey (Lund et al., 1981); for the primary somatosensory cortex in the mouse (Lorente de Nó, 1938); and for the motor cortex of the mouse (Schüz and Münster, 1985). Katz (1987) reports that layer VI pyramids in the cat visual cortex that project to the claustrum have horizontally directed local axon collaterals that arborize exclusively within layer VI and lower layer V. By contrast, most of the pyramids in cat visual cortex that project to the lateral geniculate nucleus lack horizontal collateral branches; their axons ascend and arborize within layers V and IV. Finally, the discovery by DeFelipe et al. (1986a) that terminal branches of widely separated parts of the same collateral and even of separate collaterals from the same cell can converge on a single focus of terminations is consistent with the view that the distribution of axon collaterals is orderly and with the notion that the cerebral cortex is a highly ordered structure.

III

V

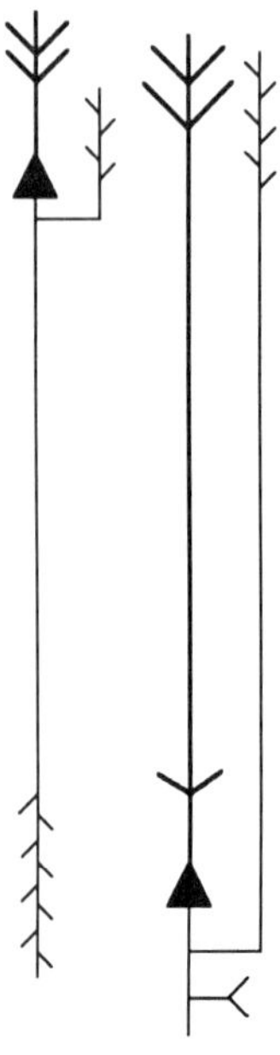

Figure 2.2. *Schematic showing projection of layer III pyramidal neurons to layer V. and of layer V pyramidal neurons to layer III. Dendrites are represented by thick lines: axons by thin ones.*

An important aspect of pyramidal cell local axon collaterals lies in their ability to serve as vertical links between the separate elements of multicellular neuronal assemblies. This feature has received increased attention in recent years with the recognition that the cerebral cortex contains vertical arrangements of neurons, that is, the functional columns. The anatomical substrate for functional columns is unclear, but it seems likely that local axonal projections that form vertical links between the different cortical laminae play a significant role in the formation of functional columns. As we shall see below, pyramidal cells are not the only cell type that contribute to intrinsic, vertical pathways; for many types of nonpyramidal neurons have local axonal projections that ramify nearly exclusively within laminae above or below the location of the parent cell body.

Laminar Relationships of Cortical Efferent Cells

Observations on the laminar location of pyramidal cell bodies labeled by the retrograde transport of substances injected into other regions of the brain have demonstrated a close relationship between the laminar location of the cell body and the site to which its axon projects (e.g., Lund et al., 1975; Wise and Jones, 1977). For instance, the somata of pyramidal cells that project to the thalamus occur in layers V and VI in the primary somatosensory cortex of the mouse (White and Hersch, 1982), rat (e.g., Jacobson and Trojanowski, 1975), and monkey (Jones and Wise, 1977); in the primary visual cortex of the rat (e.g., Peters and Saldanha, 1976), cat (Gilbert and Kelly, 1975), and monkey (Lund et al., 1975); in the primary auditory cortex of the rat (Jacobson and Trojanowski, 1975) and cat (Kelly and Wong, 1981); in the primary motor cortex of the rat (Jacobson and Trojanowski, 1975) and cat (Romagno and Maciewicz, 1975); and in the limbic cortex of the rat (Jacobson and Trojanowski, 1975) and cat (Kaitz and Robertson, 1981). These and additional data regarding the laminar location of neurons giving rise to efferent cortical pathways are depicted in Figure 2.3, which summarizes information obtained from a variety of cytoarchitectonic areas in the cerebral cortex of five species. Examination of these data shows clearly that the laminar location of pyramidal cell bodies is highly correlated with their target projection site.

As seen earlier, the dendrites and axon collaterals of pyramidal neurons also exhibit characteristic, and lamina-specific, arborization patterns. In certain instances (e.g., Katz, 1987), these patterns have shown a high correlation with target projection sites, and it would be interesting to determine whether such correlations are a general feature of the morphology of cortical pyramidal neurons. If a high degree of correlation is consistently shown between the laminar location of a pyramidal cell body, the distribution patterns of its axons and dendrites, and its target projection area, then, knowing some of these features would allow predictions to be made regarding the remainder.

Pyramidal neurons in layers II and III have axons that project to other regions of the cerebral cortex; pyramidal neurons in the deeper layers project, in addition, to subcortical regions. Thus, with respect to the cerebral cortex, the old idea that the superficial layers are receptive or associative but that the deeper layers are specialized for output (e.g., Bolton, 1910) would seem to possess some degree of validity.

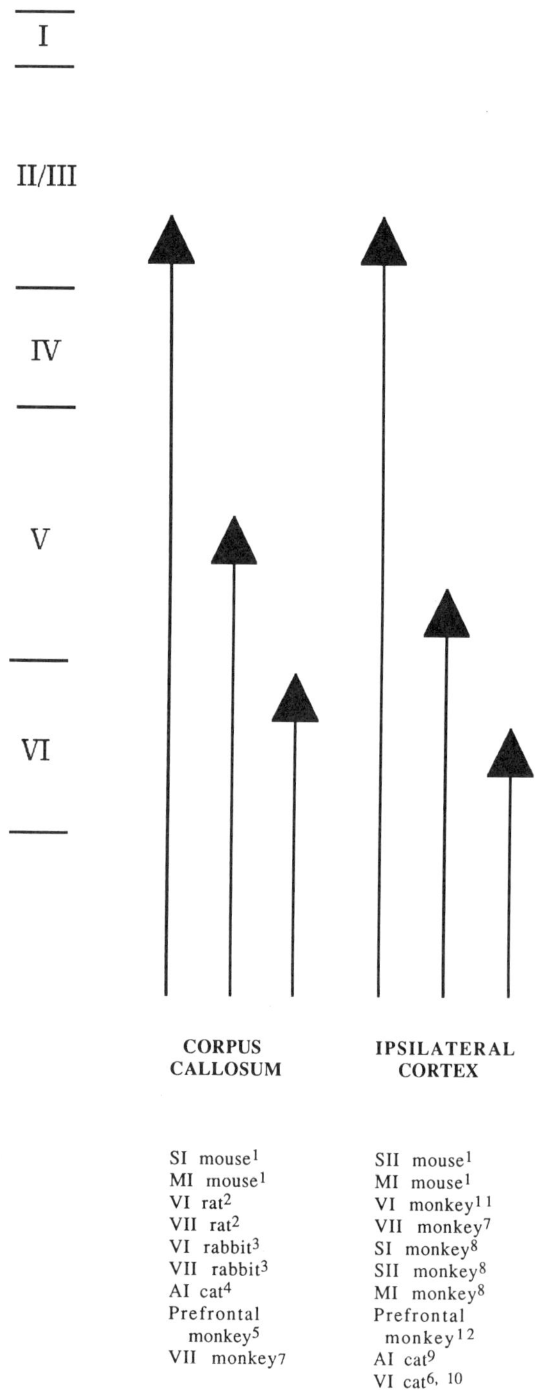

Figure 2.3. *Depiction of the laminar location of cell bodies (black triangles) belonging to pyramidal neurons listed according to their target projection area. Listed under each projection are the region of cortex and the species in which the projection cell type has been observed. Superscripts refer to the following list of references: 1, White and DeAmicis (1977); 2, Olavarria and Van Sluyters (1985); 3, Swadlow et al. (1978); 4, Kelly and Wong (1981); 5, Jacobson and Trojanowski (1977a,b); 6, Lund et al. (1979); 7, Lund et al. (1981); 8, Jones and Wise (1977); 9, Winguth and Winer (1986); 10, Gilbert and Kelly (1975); 11, Lund et al. (1975); 12, Jacobson and Trojanowski (1975); 13, Wise and Jones, (1977); 14, Peters and Saldanha (1976); 15, Romagno and Maciewicz (1975); 16, Hersch and White (1982); and 17, Jones et al. (1977).*

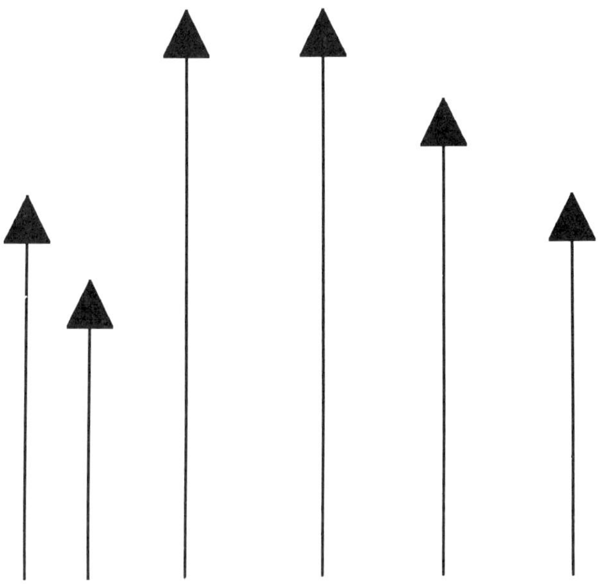

SPECIFIC THALAMIC NUCLEI	TECTUM	STRIATUM	PONS MEDULLA	SPINAL CORD
SI mouse[1]	AI cat[4]	SI mouse[16]	SI rat[13]	SI rat[13]
AI rat[12]	VI cat[10]	SI rat[13]	SI monkey[8]	SI monkey[8]
MI rat[12]	VI monkey[11]	SI monkey[17]	MI monkey[8]	MI monkey[8]
SI rat[13]	VII monkey[7]	MI monkey[17]		
VI rat[14]				
AI cat[4]				
MI cat[15]				
VI cat[10]				
VI monkey[6]				
VII monkey[7]				
SI monkey[8]				
MI monkey[8]				

Neurotransmitters of Pyramidal Cells

Tritiated glutamate and aspartate, injected into various subcortical and cortical areas, are transported retrogradely into the perikarya of pyramidal neurons (Barbaresi et al., 1987; Baughman and Gilbert, 1981; Cuénod et al., 1982; Streit and Cuénod, 1979). Indirect evidence suggests that, within the cortex, glutamate and aspartate mediate the excitatory interactions of axon collaterals belonging to pyramidal neurons (Baughman and Gilbert, 1981). Kaneko and Misuno (1988) demonstrated glutaminase, the synthetic enzyme for glutamate, in cortical pyramidal neurons. Conti et al. (1987) combined the retrograde transport of HRP with glutamate immunocytochemistry to simultaneously label efferent pathways and the neurotransmitter of pyramidal cells. Taken together, the results of these studies strongly indicate that the amino acids glutamate and aspartate are used as excitatory transmitters in local and in efferent pathways formed by axons belonging to pyramidal neurons (Streit, 1984).

Summary

A prominent, identifying feature of pyramidal cells is the presence of a vertically oriented apical dendrite. Apical dendrites typically cross several layers of cortex, and so they are well placed to receive input from the multiplicity of axonal pathways that are known to ramify within specific cortical layers. This arrangement, which allows for the integration by single neurons of a large variety of lamina-specific inputs, can be considered a crucial aspect in the design of pyramidal cells as the principal output neuron of the cerebral cortex.

Apical dendrite side branches and basal dendrites of pyramidal neurons exhibit specific spatial relationships with the cortical laminae. These relationships can be correlated with the laminar location of the parent cell body and with the target region to which its main axon projects. Each lamina of cortex contains a specific complement of axon terminals from specific sources. Thus, the study of dendrite–lamina relationships can provide insight into the kinds of synapses pyramidal cells may receive.

Pyramidal cells typically have one main axon, which projects out of the region of cortex in which the parent cell body is situated. The main axon emits several axon collaterals, which ramify within close proximity to the parent cell body and which may also ascend, descend, or travel for long horizontal distances within the cortex. Terminals belonging to the local axon collaterals of pyramidal cells are an important source of presynaptic elements in the cortex: Vertical connections between laminae are partly responsible for the formation of functional columns; horizontal connections serve to link neurons in different functional columns. Both local and distant projections of pyramidal cell axons are presumed to use the excitatory neurotransmitters aspartate and glutamate.

The specific arborization patterns of pyramidal cell axon collaterals are related to the laminar location of the parent cell body and to the terminal site to which its principal axon projects. Combining this information with that on the arborization patterns and synaptic connectivities of pyramidal cell dendrites should assist in the development of functional classification schemes for cortical pyramids. With rare exception, the arrangement within specific cortical laminae of pyramidal somata according to their efferent projections is identical in different species, an observation that provides additional confirmation that the cerebral cortices of different species are organized along similar lines.

Nonpyramidal Cells

Neurons that lack a single, dominant apical dendrite are considered to be nonpyramidal, but included within this broad category are types of neurons that differ markedly in the morphology of their cell bodies and in the shapes and patterns of distribution of their axons and dendrites. Various classification schemes have been advanced for classifying nonpyramidal cell types, but because of its simplicity, the one advanced by Feldman and Peters (1978) has proven especially useful. According to this scheme, nonpyramidal cells are classified as multipolar, bipolar, or bitufted according to the shape of their dendritic tree and as spiny, sparsely spiny, or smooth (nonspiny) according to the concentration of spines emerging from their dendrites. Multipolar cells have dendrites that emerge from multiple points on the somal surface and radiate out in all directions. Bipolar and bitufted neurons emit, respectively, either single or multiple dendritic processes from the opposite poles of cell bodies that tend to be ovoid or spindle shaped. This classification scheme is sufficient for differentiating nonpyramidal cell types according to the shapes of their dendritic trees. However, a more comprehensive scheme can be achieved by taking into account the form and distribution of the axonal ramifications of the nonpyramidal neurons.

An exhaustive presentation of all the various types of nonpyramidal neurons is beyond the scope of this work; for this the reader is referred to the general account of nonpyramidal neurons provided by Fairén et al. (1984), in which the authors have taken great pains to present a comprehensive, comparative anatomical overview of nonpyramidal cell types. Additional data on selected nonpyramidal cell types are presented in separate review chapters within the same volume in which the review of Fairén et al. appears (Vol. 1 of *The Cerebral Cortex,* Jones and Peters, eds., 1984). In the present overview of nonpyramidal cell types, we shall emphasize those types that have been observed in various areas of the cortex in different mammalian species, on the assumption that these neuronal types are the minimum necessary to support basic cortical functions (cf., the discussion of an elementary cortical pattern as purported to exist by Lorente de Nó, 1938, p. 316). Moreover, because the present work is devoted to an elucidation of cortical circuitry, we shall provide concise descriptions of only those types of nonpyramidal neurons whose morphology and synaptic connections have been investigated most intensively. The emphasis in these descriptions will be placed on those features that are directly relevant to cortical circuits.

One of the most obvious of these features, and the one on which is based our notion of the nonpyramidal neuron as a *local* circuit element, is that each nonpyramidal neuron has an axon with extensive local ramifications. Some varieties of nonpyramidal neurons have axons that project either contralaterally (e.g., in the cat visual cortex [Innocenti and Fiore, 1976] or in the cat auditory cortex [Code and Winer, 1985]) or ipsilaterally (Winguth and Winer, 1986), but as a general rule, the axonal ramifications of nonpyramidal cells are local and do not leave the area of cortex in which their parent cell body is situated. For the most part, the identity of the specific elements with which the axons of nonpyramidal neurons synapse is unknown, and so in most cases we can only speculate as to the degree of specificity with which these neurons select their targets. An implication of our proposition that the functions of the cerebral cortex are mediated primarily by circuits that are anatomically stable is that these circuits are composed of specific types of synaptic connections. Evidence that the connections formed by the local axonal ramifications of nonpyramidal cells may be highly specific with regard to the type of postsynaptic element contacted is nowhere more clear than in the synaptic connections established by the type of nonpyramidal neuron known as the *chandelier cell.*

Chandelier Cells

Chandelier cells have been so named because of the resemblance of their axonal ramifications and vertically oriented arrays of axon terminals to the branches and candles of a chandelier (Figure 2.4). Cell bodies belonging to chandelier cells have been observed in layers II through V in a variety of cortical areas and species, including the visual cortex of the mouse (Valverde, 1983), rat (e.g., Peters et al., 1982), rabbit (e.g., Müller-Paschinger et al., 1983), cat (e.g., Fairén and Valverde, 1980), monkey (e.g., in VI, Somogyi et al., 1982; e.g., in VII, Lund et al., 1981), and human (Marin-Padilla, 1987); the somatosensory cortex of

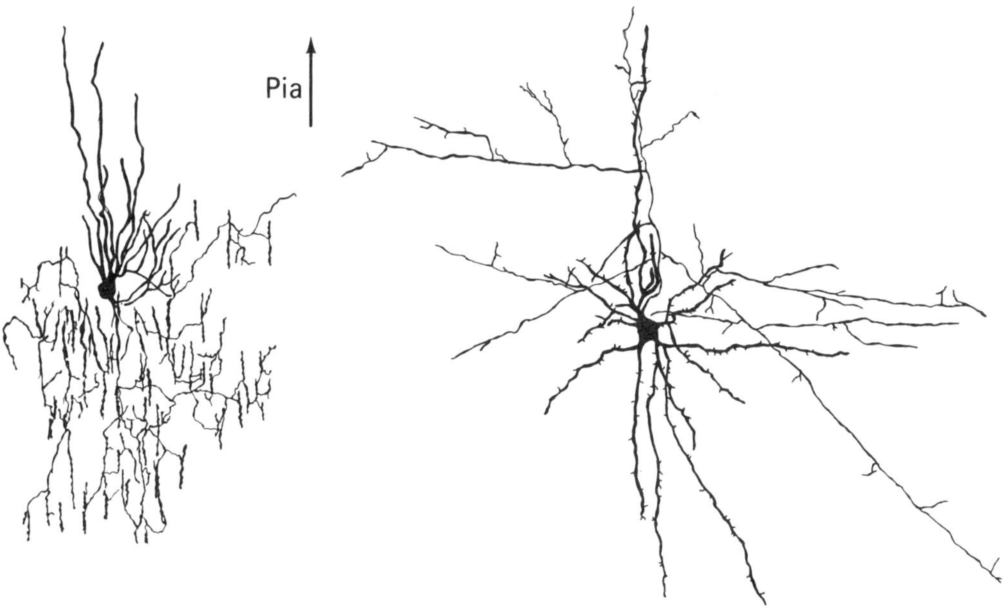

Figure 2.4. *Drawing of Golgi-impregnated neurons, at the left a chandelier cell from the frontoparietal cortex of the cat (Valverde, 1983) and at the right a basket cell from the human motor cortex (Fairén et al., 1984). Figure on left reprinted with permission by Elsevier, Amsterdam, from "A Comparative Approach to Neocortical Organization ..." by F. Valverde, p. 165 in* Ramon y Cajal's Contribution to the Neurosciences, *Grisola, S., et al., eds. (1983). Figure on right reprinted with permission of Plenum Publishing Corporation from Chap. 6 of "Non-Pyramidal Neurons" by Fairén et al., p. 239 in* Cerebral Cortex, *vol. 1, Peters, A., and Jones, E.G., eds. (1984).*

the cat (Tömböl, 1978), hedgehog (Valverde, 1983), and monkey (Jones, 1975); the auditory cortex of the mouse (Fairén et al., 1981), cat (DeCarlos et al., 1985), monkey (e.g., Szentágothai, 1975), and human (DeCarlos et al., 1987); and the temporal cortex of the human (Kisvarday et al., 1986a). The occurrence of chandelier cells in additional areas of the neocortex and in nearly every cortical layer (see Fairén et al., 1984), as well as in the cingulate (Vogt and Peters, 1981), subiculum, and piriform cortices of the rat (Somogyi et al., 1982) and in the hippocampus of the monkey (Somogyi et al., 1983b), suggests that chandelier cells are an integral component of the cerebral cortex in all mammalian species.

Chandelier cells may occur as bitufted or multipolar varieties even within a single region of the cortex. Their dendrites, which may span one or several layers of the cortex, have few branches and bear only occasional spines. Typically, the axons of chandelier cells form a profuse plexus in the vicinity of the cell body, coextensive with or somewhat above or below the distribution of the cell's dendritic tree. Some chandelier cells in the upper layers of the cortex may also possess a second, less extensive axonal ramification in the deeper layers of the cortex (e.g., Fairén and Valverde, 1980). Vertical rows of boutons belonging to chandelier cell axons form only symmetrical synapses, and these are presynaptic only to the axon initial segments of pyramidal cells (DeFelipe et al., 1985; Fairén and Valverde, 1980; Freund et al. 1983; Peters et al., 1982; Somogyi, 1977; Somogyi et al., 1979). However, as Lund (1987) observed, the presence of chandelier cell axonal boutons in layer IVC of the macaque visual cortex, where pyramidal neurons are scarce, suggests that spiny stellate cells may also receive synapses from chandelier cells.

Insight into the functional role of chandelier cells in cortical circuitry has been provided by studies showing that these neurons contain, and presumably utilize, the inhibitory neurotransmitter GABA. For example, in the visual cortex of the cat, GABAergic chandelier cells have been identified by combining Golgi impregnation with postembedding GABA immunocytochemistry (Freund et al., 1983; Somogyi et al., 1985). In addition, in the sensory-motor cortex of the monkey (DeFelipe et al., 1985) and in the visual cortex of the rat (Peters et al., 1982) and cat (Freund et al., 1983), axon terminals grouped around the initial axon segments of pyramidal neurons contain glutamic acid-decarboxylase (GAD), the rate-limiting enzyme in the synthesis of GABA.

The preferential distribution of chandelier axon terminals on axon initial segments, that is, the trigger zone for the initiation of action potentials, provides chandelier cells with the ability to exert a powerful inhibition on the outflow of activity from the postsynaptic neurons (presumably both pyramidal and spiny stellate neurons). Another type of nonpyramidal neuron that is considered to play an inhibitory role in the cerebral cortex is the basket cell.

Basket Cells

As with chandelier cells, basket cells are recognized primarily by the distribution of their axonal branches, which form nests or baskets around cell bodies and proximal dendrites belonging to pyramidal cells (Fairén et al. 1984). In his description of these pericellular arrangements, Cajal (1909–1911) noted that each is composed of branches from more than one axon collateral and that each axon collateral contributes to several pericellular nests. Cajal ascribed the origin of these nests in human tissue to neurons, which we might today identify as large multipolar cells having either smooth or sparsely spiny dendrites that often extend well above and below the layer containing the parent cell body. Subsequent work by Marin-Padilla (1969, 1970, 1972) confirmed that the pericellular nests arise from multipolar cell bodies that occur mainly in layers III and V.

The dendrites of these cells radiate in all directions, but the vertical ones predominate such that some cells have a bitufted appearance. The axons of these neurons initially ascend or descend upon leaving the cell body and then run horizontally for as much as a millimeter while contributing to pericellular nests; the presence of a long, horizontal axon is considered typical for basket cells of higher animals (e.g., Jones and Hendry, 1984, pp. 316–320). A variety of small basket cell whose dendritic trees have a somewhat different shape and whose axons ramify within more restricted areas has been identified in the cat (DeFelipe and Fairén, 1982; Fairén et al., 1984). Observations in different species that terminals presynaptic to the somata of pyramidal neurons tend to be GABAergic and to synapse at symmetrical synaptic junctions indicate that basket cells play an inhibitory role in cortical processing (in the mouse [Keller and White, 1987], rat [Ribak, 1978], and monkey [Hendrickson et al., 1981; Houser et al., 1983; and see Jones and Hendry, 1984]).

The pericellular nests formed by the axons of basket cells in the monkey (e.g., DeFelipe et al., 1986b) and human (Cajal, 1909–1911; Marin-Padilla, 1969, 1970, 1974) typically comprise several different axon collaterals. Close study of the drawings provided in these reports on Golgi-impregnated material show that the impregnation of only one or two collaterals would make the visualization of pericellular nests, and the subsequent identification of the parent cell body as a basket cell, difficult if not entirely impossible. Recognition that incomplete labeling may complicate the identification of pericellular nests undoubtedly was an important factor in the reinterpretation by Fairén et al. (1984) as basket cells of certain Golgi-impregnated neurons observed in previous investigations of a variety of cortical areas in the monkey, cat, dog, rabbit, and rodent.

The issue of whether basket cells occur in all mammals or are confined to the "higher" species has generated a good deal of controversy, and this question will be taken up later in the section on cross-species correlations of nonpyramidal neuronal types. However, from the point of view of cortical circuitry, we propose that it may be useful to include as basket cells other nonpyramidal cell types whose axons contact pyramidal cell somata whether or not true axonal baskets are formed. An example of this type would be the nonspiny multipolar cell reported by Peters and Proskauer (1980) to synapse several times at symmetrical synaptic junctions with the cell body and proximal dendrites of a layer III pyramidal neuron in the rat visual cortex.

The initial descriptions of basket cells emphasized the association of the "basket" with pyramidal neurons, and from this the impression might be gotten that basket cells synapse exclusively with the cell bodies and proximal dendrites of pyramidal neurons. However, recent evidence suggests that basket cells exhibit a variability in the types of postsynaptic elements with which they synapse. For example, counts of 241 elements postsynaptic to three basket cells in the cat visual cortex show less than half to synapse onto the cell bodies and proximal dendrites of pyramidal cells. The remainder synapse with the cell bodies and dendrites of non-pyramidal neurons and with spines, axons, and distal dendrites belonging to pyramidal neurons or to cells of undetermined origin (Freund et al., 1986; Somogyi et al., 1983a).

Direct evidence that basket cells are GABAergic has been obtained by the immunohistochemical staining of synaptic terminals belonging to basket cells that have been filled intracellularly with HRP (Kisvarday et al., 1987; Somogyi and Soltész, 1986). In addition, correlative Golgi impregnation, HRP histochemistry, and immunocytochemical studies of the monkey sensory-motor cortex (DeFelipe et al., 1986b) and the cat visual cortex (Freund et al., 1983; Martin et al., 1983) have demonstrated GABAergic neurons, which on the basis of their morphology have been interpreted as basket cells. Because the axon terminals belonging to basket cells form a large proportion of their synapses with neuronal cell bodies and proximal dendrites, it can be assumed that basket cells exert a powerful inhibition on the postsynaptic neuron by virtue of the proximity of the inhibitory synapses to the trigger zone, that is, the axon initial segment. That axons belonging to basket cells project horizontally for relatively long distances from their cell bodies suggests that these cells exert a powerful inhibition on cells in neighboring functional columns. The functional implications for columnar organization of horizontal, inhibitory interactions are discussed in detail in Chapter 5.

Vertically Oriented Neurons

The name "double bouquet" has been applied in an inconsistent fashion to neurons whose dendritic arborizations exhibit a variety of different patterns and degrees of spininess, but that may be divided into two basic types according to the distribution of their local axon collaterals and the type of synapses they form: Axons of the first type ramify in laminae situated both above and below the parent cell body, which may occur in layers II through V; terminals of these axons form *a*symmetrical synapses (cf. Fairén et al., 1984, discussing double bouquet cells with ascending and descending axon collaterals). The dendritic fields of this double bouquet type tend to be vertically elongated, that is, bitufted or bipolar in shape and coextensive with their axonal ramifications (Figure 2.5). In contrast, axons of the second type of double bouquet cell ramify for the most part within laminae well beneath their dendritic arborizations, their axon terminals form symmetrical synapses, and their cell bodies occur only within layers II and III (cf. Fairén et al., 1984, discussing double bouquet cells with mainly descending axonal arborizations). Examples of this type of double bouquet cell have been shown to contain GABA in the visual cortex of the cat (Somogyi and Cowey, 1981, 1984; Somogyi et al., 1981).

Both types of double bouquet cells have been observed in the cortices of several species including the human (Cajal, 1909–1911), monkey (Jones, 1975; Lund et al., 1981), and cat (O'Leary, 1941; Somogyi and Cowey, 1981), and at least the second type occurs in the cortex of the mouse (Lorente de Nó, 1922). With regard to interspecies differences in the morphology of the second type of double bouquet cells, Somogyi and Cowey (1984) note that from rodent to primate the axonal plexus becomes progressively more concentrated, with fewer branches and with a greater concentration of boutons on the main, radially oriented collaterals.

Bipolar cells identified in the auditory cortex of the rabbit (McMullen and Glaser, 1982), in the cingulate (Vogt and Peters, 1981) and visual cortex of the rat (Peters and Kimerer, 1981), rabbit (Globus and Scheibel (1967), cat (Fairén et al., 1984), and monkey (Peters, 1984), and in the somatosensory cortex of the mouse (Lorente de Nó, 1922) share many morphological features with double bouquet cells and could reasonably be placed in this category. In addition, like the double bouquet cells, some types of bipolar cells are presynaptic at asymmetrical synapses (Peters and Harriman, 1988; Peters and Kimerer, 1981), whereas others are presynaptic at symmetrical, presumably GABAergic, synapses (Meinecke and Peters, 1987; Peters, 1987a; Peters and Harriman, 1988).

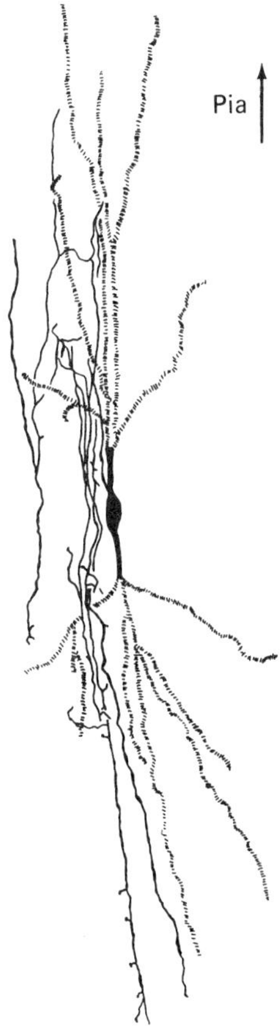

Figure 2.5. *Drawing of a Golgi-impregnated double bouquet cell from the primary visual cortex of the cat (Fairén et al., 1984). Reprinted with permission of Plenum Publishing Corporation from Chap. 6 of "Non-Pyramidal Neuron" by Fairén, et al., p. 219 in* Cerebral Cortex, *vol. 1, Peters, A., and Jones, E.G., eds. (1984).*

Spiny Stellate Cells

Because the term *stellate cell* (Sholl, 1956) has been applied to nonpyramidal cell types whose shapes may not be all that stellar, we have avoided the use of the label "stellate," referring instead to neurons with a stellar-shaped distribution of dendrites as multipolar, nonpyramidal cells. However, there is one type of spiny multipolar cell that has come to be known specifically as the *spiny stellate cell,* and we shall refer to it as such. Spiny stellate cells occur exclusively within the middle layers of the primary sensory areas of many species (see Lund, 1984). The occurrence of spiny stellate cells within laminae known to receive substantial input from the thalamus has led to the erroneous assumption that spiny stellate cells are the nearly exclusive target of specific thalamic input to the cerebral cortex (see Chapter 3). For details on the specific morphologies of spiny stellate neurons, the reader is referred to a recent review by Lund (1984). In keeping with our approach, we provide here only a capsule description of spiny stellate cell morphology with an emphasis on aspects of particular relevance to cortical circuitry.

The dendrites of spiny stellate cells are emitted at multiple points from the cell body and may have no preferred orientation. The lack of a dominant, apical dendrite is the chief characteristic by which spiny stellate cells may be distinguished from pyramidal neurons; however, in many other ways, the two cell types are identical. For instance, spiny stellate cells, like pyramidal neurons, have dendrites that bear large numbers of spines, and their axons, at least initially, descend toward the white matter and may even enter it (Figure 2.6). Both types of neuron appear similar at the fine structural level and differ from all other neuronal types in that their somata receive only symmetrical synapses. For these reasons, Lund (1984) proposed that spiny cortical neurons

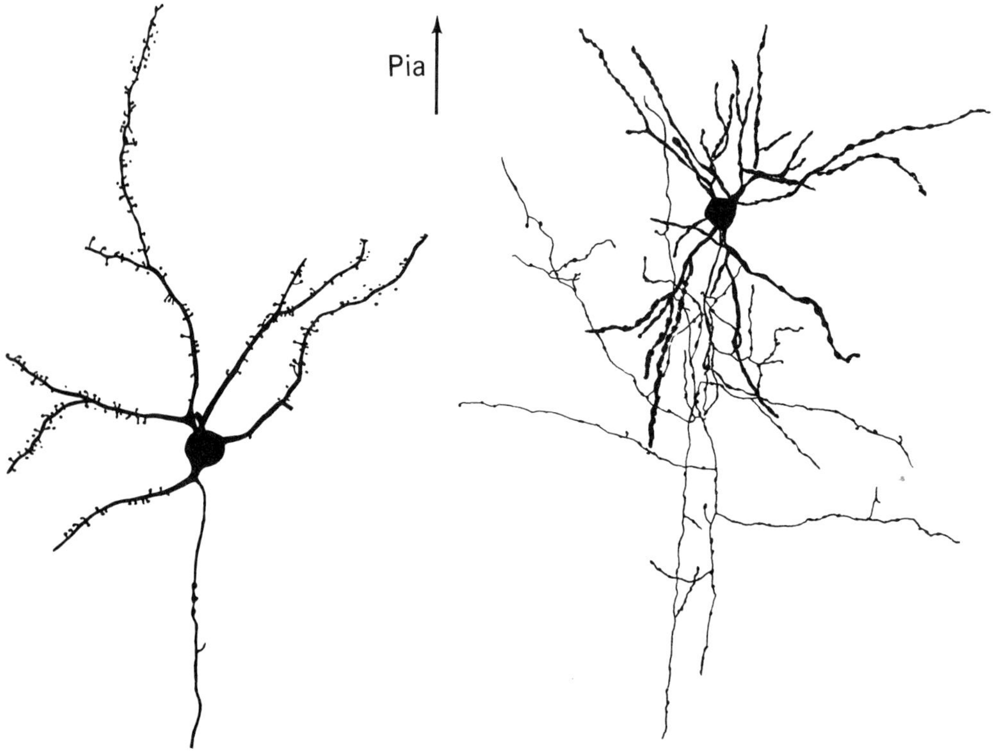

Figure 2.6. *Drawing of Golgi-impregnated neurons, at the left a spiny stellate cell from the primary somatosensory cortex of the mouse and at the right a local plexus neuron from the visual cortex of the monkey (the neuron on the right is reprinted from Peters and St. Marie, 1984). Reprinted with permission of Plenum Publishing Corporation from "Smooth and Sparsely Spinous Nonpyramidal Cells Forming Local Axonal Plexuses" by Peters, A., and Saint Marie, R.L., p. 432, in* Cerebral Cortex, *vol. 1, Peters A., and Jones, E.G., eds. (1984).*

form a continuum with the "typical" pyramidal and spiny stellate neurons as exteme types. Examples of intermediate forms would be pyramidal cells with diminutive apical dendrites (cf. Lund, 1984) and the star pyramids (see Lorente de Nó, 1922) whose dendritic distribution resembles that of spiny stellate cells with the exception that star pyramids have an apical dendrite that extends into more superficial laminae (for references on star pyramids, see Simons and Woolsey, 1984).

Certainly, from the point of view of cortical circuitry, pyramidal cells and spiny stellate cells have much in common: The distribution of synaptic types onto the dendrites and cell bodies of pyramidal and spiny stellate cells is similar (see the "Pyramidal Cells" section of Chapter 2), both cell types have axons that are presynaptic at asymmetrical synaptic junctions, where they presumably excite postsynaptic elements. In addition, both spiny stellate and pyramidal neurons show clear relationships between the laminar location of the cell body and the specific laminae and sublaminae within which their local axon collaterals ramify (e.g., Lund and Boothe, 1975). However, there are also some important differences in the general organization of the input and output relationships of these two cell types. Dendrites of spiny stellate cells are for the most part restricted to layer IV, and though they may occasionally have dendrites that enter other layers, it is clear that most of the input to them must derive from axonal pathways that terminate within this layer. The axonal output of spiny stellate cells similarly has a local quality; although some spiny stellate cells in the cat primary visual cortex project to adjacent cortical areas (Meyer and Albus, 1981), by and large, the axons of spiny stellate cells ramify within layer IV or within adjacent regions of layers III and V (Lorente de Nó, 1922, 1938; Lund, 1984). In contrast, pyramidal neurons, at least those that occur in the deeper layers, have dendritic arrangements that enable them to receive input from pathways that terminate within several cortical laminae, and the axons of pyramidal cells typically project not only locally, but also to laminae and to regions of the brain situated at considerable distance from their parent cell bodies.

Smooth and Sparsely Spiny Local Plexus Neurons

Several specific varieties of nonspiny or sparsely spiny nonpyramidal cells have been described above. Included in this group are neurons that have been termed *chandelier, basket, double bouquet,* and *bipolar cells* according to the form and distribution of their dendritic and axonal plexuses. An additional type of nonspiny or sparsely spiny neuron may be identified in the cerebral cortex. These cells, which have been referred to variously as "short-axon," "local circuit," or "Golgi type II neurons," have in common the fact that they are multipolar in shape and that their axonal ramifications include a local plexus that occurs in the immediate vicinity of the cell body and dendrites, either coextensive with them or slightly above or slightly below them (Figure 2.6). Some of these cells have, in addition, axonal branches that ramify in layers of cortex above or below the lamina containing the cell body, dendrites, and local axonal plexus.

As noted by Peters and Saint Marie (1984), whose review is a source of additional details on the morphological types and distribution of these local plexus neurons, the axons of local plexus neurons do not form easily characterized terminal arborizations, as do the axons of chandelier and basket cells. In addition, the local plexus neurons may be considered as generalized in the sense that individual cells have axons that may synapse with a wide variety of postsynaptic elements including the cell bodies, dendrites, and axon initial segments of pyramidal neurons, the cell bodies and dendrites of nonpyramidal neurons, and spines of unknown origin (LeVay, 1973; Peters and Fairén, 1978; Peters and Proskauer, 1980). Synapses with these postsynaptic elements are of the symmetrical type, and available evidence indicates that the terminals forming them are GABAergic (e.g., Hendry and Jones, 1981; Keller and White, 1987; Kisvarday et al., 1986b; Somogyi et al., 1981). Both because the axons of local plexus neurons synapse with a wide variety of postsynaptic targets and because this neuronal type appears ubiquitously within the rodent cortex, Peters and Proskauer (1980) have proposed that these neurons are the likely source in the rodent for most of the axon terminals that form symmetrical synapses.

Cross-Species Correlations of Nonpyramidal Neuronal Types

The observation that the smooth and sparsely spiny local plexus neurons synapse with a wide variety of postsynaptic targets is based primarily on findings in lower animals where this neuronal type has been most extensively examined. For this reason it is uncertain if local plexus neurons in the so-called higher species form a distinct population of neurons having generalized output relationships or whether the apparent generality in the synaptic relationships of local plexus neurons in the rodent is simply an expression of the lack of specificity reported to characterize nonpyramidal cell types in lower mammals (see below). This is an important question because findings that imply that nonpyramidal neurons are less specific with regard to their morphology and synaptic connections in lower than in higher species suggest that the cerebral cortices of lisencephalic and gyrencephalic animals are organized along basically different plans. The implications of such an arrangement for the validity of cross-species correlations of structure and function in the cerebral cortex are obvious.

Some insight into the question of the specificity of nonpyramidal neuronal types in different species can be obtained by considering certain statements by earlier workers that continue to influence our current thoughts on the organization of the cerebral cortex in light of recent findings that may require a modification of these views. For example, it has been observed that, as one ascends the phylogenetic scale, the morphology of cortical neurons becomes less uniform; that is, more specific neuronal types appear (Cajal, 1909–1911, p. 824) and the numbers of nonpyramidal neurons increases (Lorente de Nó, 1938, p. 315). In view of these statements, one might suppose that an analysis of the numbers and proportions of pyramidal and nonpyramidal cells would differ in different species. In fact, as was indicated in the introduction to Part I, this is decidedly not the case because various regions of the cerebral cortex in several species have been shown to contain identical numbers and proportions of different neuronal types (Rockel et al., 1980; Winfield et al., 1980).

It is proposed that these recent findings that emphasize the similarity of various areas of cortex in different species require a reexamination of the generally accepted notion that compared with the cortex of lower animals that of the higher animals contains a greater number of more highly specialized nonpyramidal neuronal types (see, e.g., Peters and Saint Marie, 1984, p. 434). The usual example given in support of this thesis is that neurons having long horizontal axons that form baskets of terminals around pyramidal cell bodies (i.e., true basket cells) do not occur outside the sensory and motor regions of primate cortex (cf. Jones and Hendry, 1984, p. 321). Even if this were true, it is proposed here that the appearance of true basket cells in higher species may have more to do with the size of the brain and of the neurons within it than with any real difference between the basic organization of the cerebral cortex in higher and in lower animals. To simplify discussion of this proposition, we shall presume the existence of a hypothetical basket cell in lower animals and then proceed to show why such a neuron might not be recognized readily as a basket cell.

The cerebral cortices of higher animals are thicker than those of lower animals. In accord with the argument that identical numbers of neurons underlie comparable areas of the pial surface in different species (Rockel et al., 1980), it stands to reason that neurons in higher animals display a lower packing density, that is, are farther apart from one another than neurons within the thinner cortices of lower animals. The number of pyramidal neurons contacted by individual basket cells is uncertain; however, it is clear that, to contact the same number of pyramids, the axon of a basket cell in man must necessarily travel much further than the axon must of the basket cell hypothesized for the mouse. Thus, by simply considering the distances involved in traversing the cortices of different species, it should come as no surprise that cells having long, horizontal axons (long, both in an absolute and even in a relative sense) are uncommon in lower animals. As stated previously, the presence of long horizontal axons has been and remains an important criterion for the identification of basket cells in higher animals (e.g., the review by Jones and Hendry, 1984).

A second criterion by which basket cells are identified concerns the formation by their axons of pericellular baskets around the somata of pyramidal cells. We have seen already how the identification of these baskets, and thus of the cells whose axons bear them, may be complicated by virtue of the vagaries of the Golgi method, which might fail to stain certain axonal branches that contribute to the baskets. A second factor that may complicate the identification of neurons as basket cells was alluded to previously when it was observed that each pericellular basket is composed of axon collaterals from different sources. Indeed, as DeFelipe et al. (1986b) observed in the monkey, single axonal branches contributing to pericellular baskets do not, even when they are entirely stained, take the form of a basket. Thus, the popular conception that basket cells have single axonal branches that wind around pyramidal somata to form baskets would appear to be in error: Basket cell axons contribute to baskets; they do not by themselves form them (see Figure 2.7). Presumably, it was precisely this consideration that led Jones and Hendry (1984) to reclassify as basket cells neurons identified previously in the monkey somatosensory cortex as Type I cells (Jones, 1975).

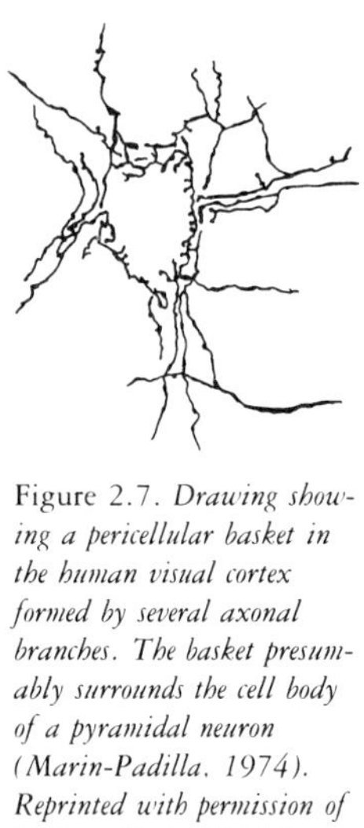

Figure 2.7. *Drawing showing a pericellular basket in the human visual cortex formed by several axonal branches. The basket presumably surrounds the cell body of a pyramidal neuron (Marin-Padilla, 1974). Reprinted with permission of Springer-Verlag Heidelberg from "Three-Dimensional Reconstruction of the Pericellular Baskets ..." by Marin-Padilla, M., in* Zeitschrift Anat. Entwicklungsges 144, *123–135 (1974).*

Problems inherent in the Golgi method, or related to the number of axons that contribute to a single basket, presumably would complicate the identification of basket cells in all species regardless of the size of the brain. Additional problems, relating specifically to the identification of basket cells, crop up in the interpretation of data obtained from the brains of lower animals as a direct result of the small brain sizes that are typical of those species commonly examined in laboratories. Cortices of smaller brains tend to be composed of neurons having smaller sizes. For instance, the somata of pyramidal cells in the visual cortex of the rat range in size from 10 to 18 μm (Peters and Kara, 1985), whereas those in the visual cortex of the cat range in size from 20 to 60 μm (Sholl, 1953). Within a single species, the larger the pyramidal cell body, the greater the number of synapses it receives (e.g., Peters and Kara, 1985), and there is every reason to believe that a similar relationship obtains across species. Thus, the smaller pyramidal cell bodies in the brains of lower animals can be expected to be postsynaptic to fewer axon terminals, and possibly to fewer axonal branches, than would the larger pyramidal cell bodies in higher species. The number of axonal branches that impinge on a pyramidal cell body in a lower animal may be sufficiently small that when viewed collectively the axonal branches do not exhibit a basket shape.

As we have emphasized previously, our main concern in the classification of nerve cells is with their synaptic connections and not with the specifics of their morphologies, and so for our purposes the identity of the target cell is considered of greater import than the morphology of the presynaptic neuron. In line with this reasoning, the interpretation of Peters and coworkers (e.g., Peters and Fairén, 1978; Peters and Regidor, 1981) that the function of basket cells in lower animals is assumed by the nonspiny or sparsely spiny multipolar neurons (i.e., the local plexus neurons) is particularly attractive. We would simply go further by stating that because of their synaptic output relationships some of the local plexus neurons in lower animals are the exact equivalent of the basket cells of higher animals and that differences in the shapes of basket cells in different species are a consequence of the different sizes and three- dimensional arrangements that characterize cortices and their component elements in different species.

Other Morphological Cell Types

The possiblity exists for further subdivision of nonpyramidal neuronal types according to the specific complement of cell surface antigens they exhibit (see Naegele et al., 1988) or by their content of neuroactive substances. These neuroactive substances include the gut-related peptides cholecystokinin (CCK), neuropeptide Y (NPY) and vasoactive intestinal polypeptide (VIP), the tachykinin substance P, and the hypothalamic peptide somatostatin, as well as GABA (for a review, see Emson and Hunt, 1984; McGeer et al., 1987). Many nonpyramidal neurons contain one or more neuroactive peptides colocalized with GABA (Gamrani et al., 1984; Hendry et al., 1984; Lin et al., 1986; Papadopoulos et al., 1987; Penny et al., 1986; Schmechel et al., 1984; Somogyi et al., 1984), and so the possibility exists of differentiating subclasses of GABAergic neurons according to the different combinations of peptides they contain (e.g., Demeulemeester et al., 1988; Jones et al., 1987). Although almost all peptide-containing neurons are also GABAergic, not all GABAergic neurons contain peptides (Hendry et al., 1984; Jones et al., 1987). For instance, chandelier and basket cells, both known to be GABAergic, do not stain for any known neuroactive peptides. In general, the physiological effects of neuroactive peptides on synaptic transmission are unclear; however, available data suggest that these substances are involved in the local modulation of synaptic activity (see Bartfai, 1985; Bartfai et al., 1986; Bloom, 1981; Lam et al., 1986).

Although various morphological cell types have been described in the preceding discussion, there are many cell types that have not been mentioned. Among these are the Cajal-Retzius neurons of layer I, whose horizontally oriented dendrites and axons would seem well placed to tie together regions of cortex that lie beneath them (see review by Marin-Padilla, 1984), and a variety of morphological types that have been identified in other layers (e.g., Lund, 1987). Discussion of these neurons has been omitted for several reasons: (1) Extensive reviews of neuronal types already exist (e.g., Fairén et al., 1984; and see other chapters in that volume), and little could be gained by repeating these efforts; and (2) the present work focuses on cortical circuitry, and in many instances, although specific cell types have been identified according to various morphological criteria, little is known of the circuitry in which they are involved. For our purposes there is scant reason to mention these neuronal types other than to remind the reader that they exist so that they can be incorporated into the general scheme of cortical circuitry as additional information regarding their synaptic connections becomes available. Moreover, inherent in our approach is the belief that the circuits within which a neuron participates are of critical importance to cortical function, whereas other morphological features such as size and shape are useful only insofar as these parameters can be correlated with cortical circuitry.

In any review of the literature concerning morphological types of neurons (e.g., Fairén et al., 1984; Feldman, 1984), it soon becomes clear that the shape of cells has consistently played a central role in the classification of neuronal types. Considerations of shape in the classification of neurons would seem justified to the extent that the shape of a neuron, including the distribution of its dendrites and axonal branches, governs the kinds of synaptic relationships in which a neuron may participate. However, three-dimensional arrangements may exist that affect neuronal shape but that have little bearing on the kinds of connections the neurons may form. An example of this phenomenon is the different shapes exhibited by nonpyramidal neurons in layer IV of the somatosensory cortex of the mouse. In this region, nonpyramidal neurons are arranged three- dimensionally to form multicellular units termed *barrels* whose sides contain relatively high concentrations of neuronal cell bodies (Woolsey and Van der Loos, 1970). The majority of the nonpyramidal neurons a barrel comprises have their dendritic arborizations within the same barrel in which their cell bodies occur (Woolsey et al., 1975). Neurons whose somata occur in the centers of barrels are stellate shaped, whereas those with somata in the sides of barrels exhibit a marked asymmetry in the distribution of their dendrites, for these are projected toward the barrel center. Both of these morphological types have access to the same elements within the barrel, and at least as regards their input, it would make no sense to classify them as separate neuronal types. One wonders how many of the morphological cell types that have been described to date owe their separate existence to similar, but as yet unrecognized, quirks in the three-dimensional arrangements of neuronal elements.

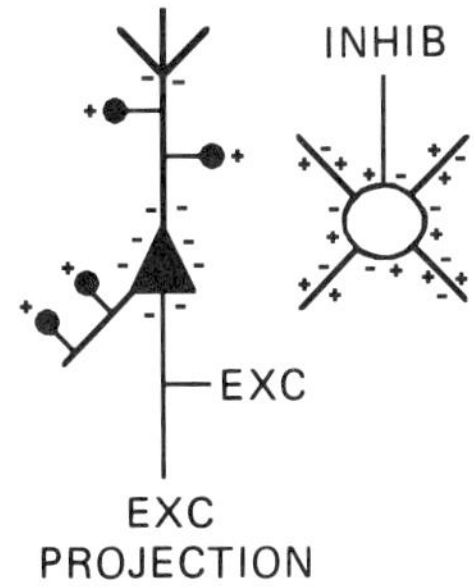

Figure 2.8. *A simplification illustrating the two basic neuronal types within the cerebral cortex: One is excitatory, and the other, inhibitory. Anatomical features considered to have important consequences for the physiology of these neurons are that the excitatory neurons have spines and the inhibitory ones generally do not; the excitatory neurons receive asymmetrical presumed excitatory, synapses (+) onto their spines and symmetrical, presumed inhibitory, synapses (−) onto their cell bodies and dendritic shafts; and the inhibitory neurons receive both types of synapse on all parts of their cell bodies and dendrites.*

The arrangements shown here are ubiquitous in the cerebral cortex and also occur in many other parts of the brain. We conclude that this figure outlines basic features that ought to characterize the components of any neuronal model whose aim is to replicate brain function. For a guide as to how these components might be connected in a neuronal model, the reader is referred to the rules presented at the ends of Chapters 3 and 6.

Summary

Nonpyramidal cells come in many morphological varieties, which are identified primarily by the distribution patterns of their axons and dendrites. The potential for further subdivision of these morphological types exists by taking into account the various combinations of peptides and amino acids they contain. Many types of nonpyramidal neurons contain the inhibitory neurotransmitter GABA and one or more of a variety of peptides.

The axonal ramifications of nonpyramidal cells rarely leave the area of cortex in which their parent cell body is situated. Therefore, from a connectional standpoint, nonpyramidal neurons have a role that contrasts sharply with that of pyramidal neurons: Both types have axons that provide synapses locally, but the pyramidal neurons participate also in long-distance connections that link the different areas of the cerebral cortex with each other and with other regions of the brain.

In general, nonpyramidal and pyramidal neurons also differ in the spatial distribution of various types of synapses they receive. For example, the cell bodies of pyramidal neurons form only symmetrical synapses, whereas those of most types of nonpyramidal neurons receive asymmetrical synapses as well. These different patterns of synapses are depicted in Figure 2.8, which is an attempt to identify those features of cortical neurons that are most relevant to the elaboration of neuronal network models. The basic types of cortical neurons are labeled excitatory and inhibitory, instead of pyramidal and nonpyramidal, to emphasize the functional aspects of cortical neurons rather than their morphological types. For instance, the spiny stellate cell, a nonpyramidal neuron, has an axon that forms asymmetrical synapses and thus would be included, with pyramidal neurons, in the "excitatory" category.

Summary of Chapter 2

This chapter has dealt with the basic neuronal elements that participate in cortical circuitry primarily on the basis of information obtained with the light microscope. The natural tendency when interpreting observations made with the light microscope is to presume an increased opportunity for synaptic interaction between axons, and dendrites and cell bodies, when these neuronal elements coexist within close proximity, and to presume a decreased opportunity for synaptic interaction when these elements are spatially separate (e.g., Bodian, 1952). For instance, synaptic circuits in the cortex that involve different populations of neurons have been proposed on the basis of the coexistence within a single cortical lamina of axon terminals belonging to one population and dendritic branches of the other (e.g., Lund and Boothe, 1975; Sholl, 1956). In like manner, the apposition of basket cell axons to the somata of cortical pyramids has been interpreted to indicate a synaptic connection between these two elements. In the course of time, the basket cell/pyramid connection, and many other synapses whose existence has been supposed on the basis of light microscopic observations, have been confirmed by direct observations made with the electron microscope (for the basket cell, cf. Cajal, 1909–1911, pp. 556–558, with, e.g., Somogyi et al., 1983a).

A particularly meaningful approach to the light microscopic analysis of synaptic organization in the cerebral cortex was adopted by Sholl (1956), who attempted to simplify observations on cortical synaptic connectivity by expressing them in statistical form. Sholl argued that the statistical treatment of aggregates of neurons, taking into account various quantitative data such as the numbers, sizes, and packing density of neuronal somata, and the lengths and distribution patterns of axons and dendrites is the most appropriate method to adopt for a system composed of a large number of components whose interactions can only be described in terms of probability. For additional insights into the organization and function of the cerebral cortex based on statistical analyses of its myriad components, and taking into account recently identified features of cortical organization, see Braitenberg (1977a,b; 1978, 1985).

A distinct advantage of the light microscopic approach over other methods for elucidating cortical circuitry is the fact that large numbers of putative synaptic contacts can be identified and plotted precisely along postsynaptic elements (e.g., Harris and Woolsey, 1983). Using this approach should facilitate comparing patterns of synaptic input with the results of physiological recording. This is a necessary step on the way toward modeling the activity of neurons in response to specific patterns of synaptic inputs.

In some instances, results obtained by light microscopy have been misinterpreted because the light microscope lacks the level of resolution necessary to visualize synapses. In fact, as we shall see in Part III, interpretations based on light microscopic studies have led to the formulation of elaborate theories of cortical function that have been based on misapprehensions of the circuitry actually present. The problem is that observations made with the light microscope only imply the existence of a synaptic contact between specific neuronal elements—the synapse may not actually be formed. Indeed, Peters and Proskauer (1980) cite several instances in which predictions of synapses made on the basis of light microscopic observations were discounted subsequently on the basis of evidence obtained with the electron microscope. Moreover, at the level of resolution offered by the light microscope, it is impossible to determine how many synapses are formed by an axonal bouton. Some boutons do not form synapses, whereas others form multiple contacts and none of this information is evident with the light microscope. At present, the electron microscope is the only instrument that permits synaptic connections to be visualized directly and reliably, a feature that has made electron microscopy the most routinely applied approach for the unequivocal identification of synaptic connections.

3 Synaptic Connections Between Identified Elements

The investigation of synaptic relationships in the cerebral cortex has recently entered a most productive era with the introduction and increasing application of combined light and electron microscopic methods that for the first time have enabled the synaptic relationships of identified neuronal elements to be determined reliably and on a regular basis. The methods currently employed combine the identification with the light micoscope of neurons impregnated by the Golgi method (e.g., Blackstad, 1965; Fairén et al., 1977) or labeled by the retrograde transport (e.g., White et al., 1980) or intracellular injection (e.g., McGuire et al., 1984) of horseradish peroxidase (HRP) followed by the analysis of their synaptic connections with the electron microscope.

Obtaining data on synaptic connectivity with the electron microscope is a time-consuming process that can be rendered far more efficient by labeling both the pre- and postsynaptic elements. Presynaptic terminals from sources extrinsic to the cortex may be labeled by anterograde transport methods or by lesion-induced anterograde degeneration. For a variety of technical reasons, labeling terminals from extrinsic sources is far more easily accomplished than the labeling of terminals from intrinsic sources (see the "Synapses of Extrinsic Afferents with Cortical Neurons" section below.) Because of this, much more information is available regarding the circuitry of terminals from extrinsic than from intrinsic sources.

Despite the relatively incomplete picture available for synaptic re-

---

Recent research efforts of E.L. White and E. Elhanany were supported by grants from The National Institutes of Health (Grant NS 20149) and from The U.S.–Israel Binational Science Foundation (Grant 86-00041).

lationships involving terminals from intrinsic sources, the importance of these relationships for cortical processing should not be underestimated. Indeed, in many areas of the cerebral cortex, synapses of local origin may account for as many as 70% of the synapses present (e.g., Gruner et al., 1974; Szentágothai, 1965), and it is difficult to imagine that the qualities that set the cerebral cortex apart from other regions of the brain are not due in large part to the richness of the intrinsic circuits within it.

The essence of the study of cortical circuitry involves determining the types, numbers, and spatial distribution of synapses linking neuronal elements, and so in this and in subsequent sections, certain quantitative aspects of cortical synapses will be considered in detail. It will become apparent that most information on the quantitative aspects of cortical synapses has been acquired only recently and that the results are limited to only a few cortical areas in even fewer species.

Considerable evidence has been presented to support the contention that the cerebral cortex is organized along the same basic lines in different species, but whether this similarity extends to the precise composition of the neuropil within corresponding laminae is as yet unknown. This information is critically important because interpretations based on comparisons of quantitative data obtained from different species, and also from different layers of the same species, may depend to a large extent on whether the composition of the neuropil in the different regions is similar. For example, finding that the output of a neuron in one region (or species) is mainly to spines but in another mainly to dendritic shafts might mean that the neuron is programmed to form different output relationships in different regions (or species); however, it is also possible that the output relationships merely reflect the composition of the neuropil in the regions examined. This is but one example of why it is useful to include information on the composition of the neuropil in any quantitative study of synaptic connections, a topic to be taken up in greater depth in Chapter 7.

Intrinsic Connections of Pyramidal Neurons

Terminals belonging to the local axon collaterals of pyramidal neurons have been examined with the electron microscope in several areas of the cerebral cortex in different species (e.g., in the visual cortex of the rat [Parnavelas et al., 1977; Somogyi, 1978], cat [Gabbot et al., 1987; Kisvarday et al., 1986c; McGuire et al., 1984], and monkey [LeVay, 1973]; the somatosensory cortex of the mouse [White et al., 1980; White and Hersch, 1981, 1982; White and Keller, 1987] and monkey [Winfield et al., 1981]; and the piriform cortex of the opossum [Haberly and Presto, 1986]). These terminals are presynaptic only at asymmetrical synaptic junctions mainly to dendritic spines and shafts and rarely to neuronal cell bodies. Reports on the proportions of synapses formed with spines versus dendrites have been inconsistent, ranging from mostly dendrites or mostly spines being contacted to approximately equal proportions of spines and dendrites.

Local Projections of Superficial Pyramids

At present, there are several reports that provide quantitative data on the synaptic connections of local axon collaterals belonging to pyramidal neurons having somata in the upper layers of the cortex. These reports include observations made in the somatosensory cortex of the mouse (White and Hersch, 1981) and monkey (Winfield et al., 1981), the visual cortex of the rat (Somogyi, 1978) and cat (Kisvarday et al., 1986c), and the piriform cortex of the opossum (Haberly and Presto, 1986). From an evolutionary perspective, the piriform cortex is the most primitive region of the cerebral cortex (Nauta and Karten, 1970). Accordingly, its neuronal composition and laminar organization are simpler than that of the neocortex that forms the basis of most of the present discussion.

Observations common to each of the reports cited above are that the main trunks giving rise to the local axon collaterals of superficial pyramidal cells are nearly always myelinated either before or after giving off their collateral branches and that the collateral branches, which are unmyelinated, form synapses only of the asymmetrical type. In rare instances the collaterals synapse onto cell bodies presumed to belong to nonpyramidal neurons (Kisvarday et al., 1986c). Synapses onto dendritic shafts and spines are far more common, and these occur as two basic arrangements: The local axon collaterals either form most of their synapses with spines (Figure 3.1, bars 1–4) or, in contrast, form roughly equal proportions of their synapses with spines and dendritic shafts (Figure 3.1, bars 5–8).

Kisvarday et al. (1986c) examined pyramidal neurons in the primary visual cortex and compared the synaptic relations of their axon collaterals near the parent cell body in layer III with synaptic relations of other collaterals situated directly underneath, in layer V. In addition, they examined ramifications of the same axons in both layers III and V at some distance away from the region of the cell body. Terminals of the local axon collaterals made similar proportions of their synapses with spines and dendritic shafts in each region examined, suggesting that the local axon collaterals of pyramidal neurons consistently form the same proportion of synapses with dendritic shafts versus spines throughout all parts of their ramification.

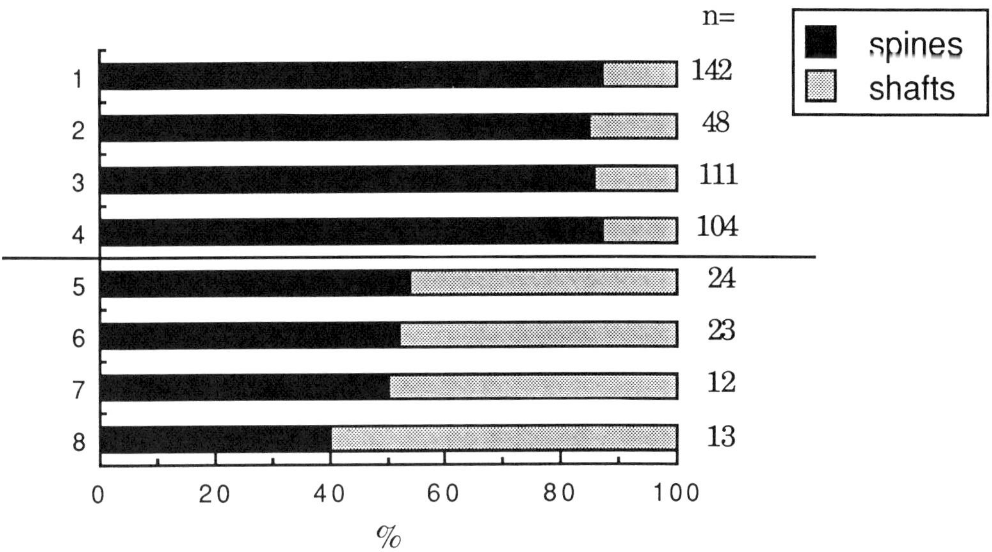

Figure 3.1. *Bars representing percentages of synapses onto spines and dendritic shafts made by the axon collaterals of superficial pyramidal neurons. All synapses are of the asymmetrical type.* n = *total number of synapses. Bar 1, cat VI cortex, two layer III pyramids, collaterals in layer III (Kisvarday et al., 1986c).; bar 2, cat VI cortex, two layer III pyramids, collaterals in layer V (Kisvarday et al., 1986c); bar 3, mouse SmI cortex, unknown number of presumed layer III pyramids, collaterals in layer III, (White and Elhanany, unpublished); bar 4, mouse SmI cortex, unknown number of presumed layer III pyramids, collaterals in layer V, (White and Elhanany, unpublished); bar 5, opossum piriform* cortex, two layer II pyramids, collaterals in layer III (Haberly and Presto, 1986); bar 6, two layer IV pyramids**, rat VI cortex, collaterals in layer IV (Somogyi, 1978); bar 7, two layer III pyramids, mouse SmI cortex, collaterals in layer IV (White and Hersch, 1981); and bar 8, one layer III pyramid, monkey SI cortex, collaterals in layers II and III (Winfield et al., 1981).*

*Piriform cortex has but three layers, and so it is difficult to compare the laminar arrangement of neurons and synapses in this region of cortex with those occurring in the six-layered neocortex. Data include three synapses onto shafts and one onto a spine that occurred deep in layer III, a region that could be considered as equivalent to the deeper regions of neocortex.
**Includes one cell identified by the authors as a spiny stellate neuron; however, this cell possesses morphological features consistent with our identification of it as a superficial pyramidal neuron.

Data on the composition of the neuropil in the different regions were not provided, and so it cannot be determined if these results reflect the proportions of axospinous and axodendritic synapses in the neuropil of the different regions. Evidence to the contrary, indicating that the output of single pyramidal cells may vary with cortical depth, has been provided by Haberly and Presto (1986). They observed in piriform cortex that, in the superficial regions of layer III, most synapses of pyramidal cell axon collaterals were with dendritic spines (6 spine, 1 shaft), whereas in mid–layer III the collaterals formed approximately equal numbers of shaft and spine synapses (6 spine, 7 shaft). In the deeper regions of layer III, the collaterals tended to synapse preferentially with dendritic shafts (1 spine, 3 shaft). Similarly, in a report on synapses made by the axon collaterals of layer V pyramidal neurons in areas 17 and 18 of cat visual cortex, Gabbott et al. (1987) observed that the percentage of synapses formed onto dendrites as opposed to spines increases markedly from a value of zero onto dendrites in upper layer IV to as high as 35% onto dendrites in layer VI.

Additional data suggesting that the output relationships of pyramidal cells may differ in different laminae are provided by a comparison of results from White and Hersch (1981) with those of an ongoing study (White and Elhanany, personal observations). White and Hersch (1981, Fig. 4) report on synapses made in layer IV of the primary somatosensory area by the local axon collaterals of two layer III pyramidal neurons labeled by the retrograde transport of HRP injected into ipsilateral motor cortex. Half of the synapses are with spines, the other half are with nonspiny dendritic shafts. White and Elhanany's study is an examination of synapses in layers III and V, of the same region of the mouse cortex, made by axon collaterals belonging to pyramidal cells that were labeled precisely as they were by White and Hersch (1981). Preliminary findings show that, in both layers III and V, more than 85% of the synapses are formed with spines, suggesting that the proportions of the different postsynaptic elements contacted by the axons of superficial pyramidal cells may be very different in layers III and V from IV. Some of the synapses examined by White and Elhanany are made by collaterals of pyramidal neurons whose somata occur in layer V, and so it is not accurate to consider all the findings as pertaining to layer III pyramids. Moreover, these findings include data from an unknown number of labeled pyramidal cells and as such are not entirely comparable to those reported by White and Hersch (1981) and Kisvarday et al. (1986c), who examined the output of specific, identified pyramidal cells.

In view of the above discussion, it is difficult to determine whether the local axon collaterals of cortical pyramidal neurons typically exhibit similar synaptic relationships within all parts of their ramification or whether the synaptic relationships of these collaterals are different in different laminae. Conceivably, some neuronal types display one of these patterns, and other types display the other; that different collaterals belonging to a single neuron may display both patterns cannot be excluded (cf. differences in the anatomical organization of different branches belonging to a single thalamocortical afferent, Landry et al., 1987). One feature common to all the findings cited in this context is that, in all cortical laminae examined, the local output of superficial pyramidal neurons was always to at least two neuronal types, one having spines and the other lacking them. This observation serves to emphasize the point that the axonal outflow from a single source may be to a number of entirely different neuronal elements. Thus, when considering cortical circuitry, we must think not only in terms of chains of interconnected neurons, one synapsing with the next, but rather, in a broader sense, of more complex interactions between networks of interconnected neurons, a theme to be more fully developed in Part III.

Neurons postsynaptic to the axon collaterals of superficial pyramidal neurons were identified by similar methods and criteria in each of the studies cited in Figure 3.1. For dendrites these criteria included the presence or lack of spines, the overall pattern of synapses received by the dendrite, and the shape of the dendrite and of the cell body, from which it issued. In one instance, a dendrite postsynaptic to the terminal of a pyramidal cell axon collateral was traced back to its parent cell body, which was interpreted to belong to a large, presumably nonspiny, nonpyramidal cell (Somogyi, 1978). Other dendrites contacted by labeled axon terminals were not traced back to their parent cell bodies, but tentative identifications of them were made on the basis of the synaptic patterns they formed. For example, in each of the studies cited, some dendrites postsynaptic to pyramidal cell axon collaterals lacked spines and were postsynaptic to numerous unlabeled axon terminals that made asymmetrical synapses. These are features typical of dendrites belonging to nonspiny or sparsely spiny nonpyramidal cells (e.g., in the rat [Parnavelas et al., 1977], cat [Colonnier, 1968; LeVay, 1973], mouse [White et al., 1984], opossum [Haberly and Feig, 1983], and monkey [Sloper et al., 1979]).

Some nonspiny dendrites receiving synapses from the axon collaterals of superficial pyramids are immunoreactive for GABA (Kisvarday et al., 1986c), a substance that to date has been identified only within nonpyramidal neuronal types (e.g., Houser et al., 1984, pp. 79–84; and see Chapter 6). Certain of the nonspiny or sparsely spiny nonpyramidal cells described earlier, including the basket, chandelier, double bouquet, and local plexus cells, may be GABAergic (see Chapter 2), and all are known to have dendrites that receive large numbers of asymmetrical synapses. Thus, from the point of view of cortical circuits, it is worth considering that any of these GABAergic, nonpyramidal neuronal types may receive input directly from the axon collaterals of superficial pyramidal neurons.

The identity of the spines that synapse with the local axon collaterals of superficial cortical pyramids is somewhat more difficult to determine. Candidates for the origin of these spines include spiny stellate cells, other types of spiny nonpyramidal cells (e.g., spiny bitufted cells [Hornung and Garey, 1981; White and Hersch, 1981]) and pyramidal neurons. Haberly and Presto (1986) provide evidence indicating that, at least in the piriform cortex, the most likely source for the spines contacted by the axon collaterals of pyramidal neurons are other pyramidal neurons.

This conclusion is based on two observations: (1) that the spines in question resemble those of pyramidal neurons, but are very different from those of most types of nonpyramidal neurons in the piriform cortex (Haberly, 1983), and (2) that dendrites belonging to those types of nonpyramidal neurons in piriform cortex that do have spines resembling those of pyramidal cells are not present in most of the regions examined (Haberly and Presto, 1986).

Some of the synapses made by the axon collaterals of superficial pyramidal cells examined by Kisvarday et al. (1986c) were onto spines that issued from dendrites whose fine structural characteristics resembled those of pyramidal cells. For this reason and because dendrites of spiny nonpyramidal neurons have been but rarely observed in Golgi preparations within the laminae examined by Kisvarday et al. (1986c), these authors conclude that the majority, if not all, of the spines postsynaptic to the local axon collaterals of superficial pyramidal neurons derive from other pyramidal neurons.

Conclusive evidence that axon collaterals of superficial pyramidal cells form synapses with other pyramidal cells is provided by White and Hersch (1981), who examined the synaptic connections of superficial pyramidal cells that had been labeled by the retrograde transport of HRP. They depict a synapse between two HRP-labeled processes, an axon collateral of a superficial pyramidal neuron, and a spine belonging to a different pyramidal neuron. The laminar location of the cell body belonging to the postsynaptic pyramidal neuron was not determined.

In summary, the impression obtained from examination of Golgi preparations is that, except within layer IV, where dendrites of spiny stellate cells abound, there are few spines that belong to nonpyramidal cell types. For this reason, it seems reasonable that, outside layer IV, most spines and spiny dendritic shafts belong to pyramidal neurons; within layer IV, a large proportion no doubt belong to spiny stellate cells. Therefore, outside of layer IV, spines postsynaptic to the local axon collaterals of superficial pyramidal cells most likely belong to other pyramidal neurons. Below layer IV, these spines would belong to deep pyramidal neurons; above layer IV, these spines may belong to both superficial and to deep pyramidal neurons. Layer IV contains spines belonging to superficial and to deep pyramidal neurons and, in addition, to nonspiny stellate cells; all these spines are in a position to receive input from the local axon collaterals of superficial pyramidal neurons.

The data discussed thus far in this section are considered to represent two basic kinds of circuits involving the local axon collaterals of pyramidal neurons having somata in layers II–IV. In each circuit the pyramidal neurons are presumed to excite their targets, a contention in accord with results obtained by physiological recording (c.f., Ferster and Lindström,

1985; see Chapter 5). One circuit involves as postsynaptic targets smooth and possibly sparsely spiny nonpyramidal neurons (Figure 3.2). Smooth and sparsely spiny nonpyramidal neurons are considered to be the main source of the GABAergic, symmetrical synapses that impinge on the cell bodies and dendritic shafts of cortical pyramids, and so in this circuit we have the anatomical substrates for both feedback and feedforward inhibition (e.g., see Chapter 2). The second circuit involves synapses from superficial pyramidal cells onto spiny neurons (Figure 3.2); above and below layer IV, these spiny neurons are in all likelihood other pyramidal neurons.

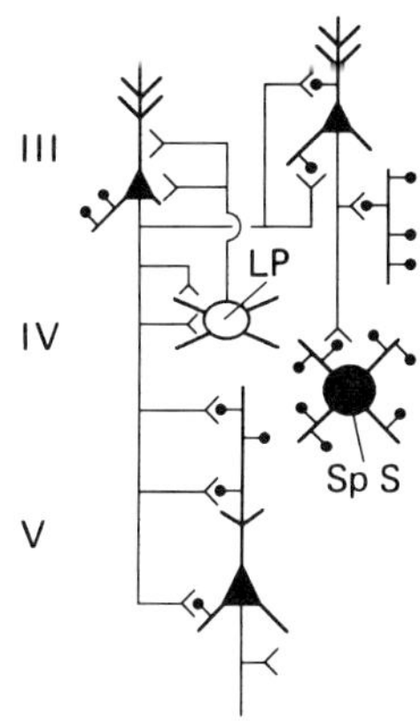

Figure 3.2. *Schematic illustration of the local projection of superficial pyramidal neurons to other layer III pyramids, to layer V pyramids, to spiny stellate cells (SpS), and to local plexus neurons (LP). A feedback projection from the local plexus neuron is also shown. Vertical spiny dendrite at the right represents the apical dendrites of pyramidal neurons whose cell bodies occur in the deeper layers of the cortex.*

The existence of a feedforward excitatory circuit from layer III to layer V pyramids has been postulated to account for the increased complexity of the receptive fields of neurons in the lower layers of the cortex with respect to neurons in the upper layers (Hubel and Wiesel, 1968, 1977; Simons, 1978, 1985). The possible role of feedforward excitatory circuits in the formation of the receptive fields of cortical neurons is considered at length in Part II in a discussion of the relationships between anatomically defined circuits and the response properties of cortical neurons.

Local Projections of Deep Pyramids

The synaptic connections of axons belonging to pyramidal cells whose somata occur in layers V and VI of the cortex have also been examined; and like those made by axons of pyramidal neurons in the more superficial layers of the cortex, their synapses are also only of the asymmetrical type (e.g., White and Hersch, 1982).

There are three studies that have investigated in depth the synaptic connections of axon collaterals belonging to layer V and VI pyramidal neurons. The earliest of these reports on the synaptic connections of local axon collaterals belonging to two pyramidal neurons whose somata occur in layer VI of the cat primary visual cortex; these neurons were stained by the intracellular injection of HRP (McGuire et al., 1984). In this study, the examination of labeled axon terminals was restricted to the region of layer IV situated above the parent cell body. The second report is of synapses made by the local axon collaterals of two layer V pyramidal cells, again in the cat primary visual cortex and again labeled by the intracellular injection of HRP (Gabbott et al., 1987); but in this study, terminals in layers IV–VI of area 17 and also in area 18 were examined. The most recent work focuses on synapses made in layers IV–VI of the primary somatosensory cortex of the mouse by local axon collaterals belonging to an unknown number of layer V and VI pyramidal neurons that were labeled by the retrograde transport of HRP injected into the ipsilateral thalamus (White and Keller, 1987).

Findings of McGuire et al. (1984) and White and Keller (1987) are in agreement in that virtually all the boutons belonging to the local axon collaterals of these deep pyramids form only a single asymmetrical synapse each, and most of the synapses formed are with dendritic shafts, the remainder with dendritic spines. However, the findings of these two studies differ in several important respects. For instance, terminals of axon collaterals belonging to pyramidal cells examined by White and Keller (1987) form nearly all of their synapses with the shafts of nonspiny dendrites, whereas those examined by McGuire et al. (1984) form a comparatively large proportion of their synapses with spines or with the shafts of spiny dendrites. Gabbott et al. (1987) also report that the local axon collaterals of deep pyramidal neurons form only asymmetrical synapses, but in contrast with the two other studies, they report that a relatively high proportion of collateral synapses are made with spines. They observed, in addition, that individual boutons can form multiple axospinous synapses; the average number of spines contacted by a single bouton was 1.9 for one cell and 1.4 for the other.

For comparison, the results of these analyses are set out in Figure 3.3, which shows the percentages of synapses formed onto spines and dendritic shafts. In two of the studies (McGuire et al., 1984; White and Keller, 1987), information was provided on the numbers of axodendritic (shaft) synapses made with dendrites belonging to different types of neurons; these results are illustrated in Figure 3.4.

The data presented in Figures 3.3 and 3.4 indicate that terminals belonging to the local axon collaterals of corticothalamic projection cells in the mouse primary somatosensory cortex (area 3; White and Keller, 1987) synapse mainly with dendritic shafts and that most of these dendrites belong to nonspiny neurons. By comparison, layer Vb and layer VI neurons in the cat primary visual cortex (area 17) show a greater tendency to synapse onto spines and, at least in the case of layer VI cells, onto dendrites that belong to spiny neurons (McGuire et al., 1984; Gabbott et al., 1987).

Part of the reason for these contrasting synaptic patterns may be related to differences in the methodological approaches employed to collect the data or to differences in the composition of the neuropil in the different species or areas of cortex examined. Such circumstances, however, are inadequate to explain the striking differences reported for the local output relationships of layer Vb (Gabbott et al., 1987) versus layer VI (McGuire et al., 1984) pyramids in layer IV of the cat visual cortex. An alternative explanation is that differences in the local output relationships of deep pyramidal neurons are related to differences in the kinds of pyramidal neurons examined. For instance, it is entirely conceivable that these three studies focus on three different populations of projection neurons: White and Keller (1987) examined terminals belonging to unequivocally

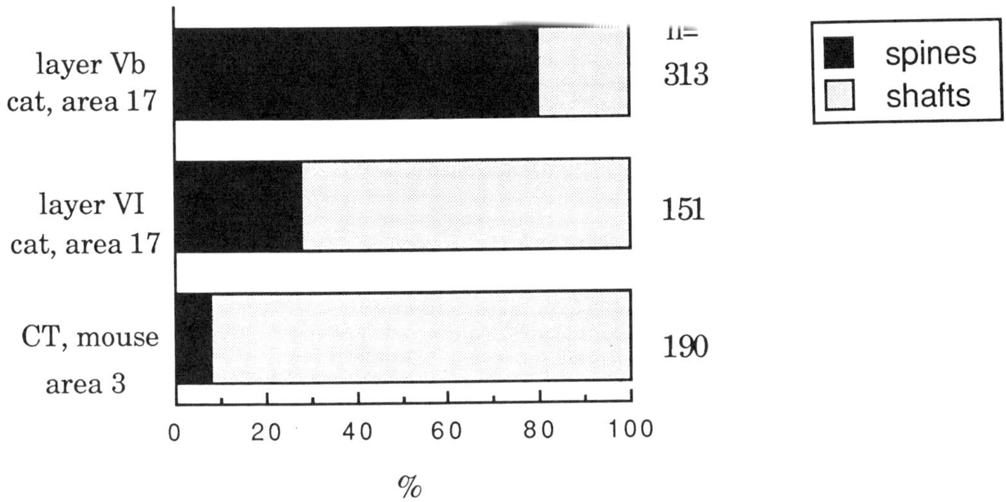

Figure 3.3. *Graph showing the percentages of synapses formed by the axon collaterals of deep pyramidal neurons onto spines and dendritic shafts. All synapses are of the asymmetrical type. Numbers of synapses appear at right of each bar (n).*

Layer Vb, cat, area 17, includes synapses made in layers IV–VI by two layer Vb pyramidal cells in the cat primary visual cortex; the authors propose that these neurons project to the tectum (Gabbott et al., 1987).

Layer VI, cat, area 17, includes synapses made in layer IV by two layer VI pyramidal cells in the cat primary visual cortex; the projection sites of these cells are unknown (McGuire et al., 1984).

CT, mouse, area 3, includes synapses in layers IV–VI made by an unknown number of corticothalamic (CT) projection cells whose somata occur either in lower layer V or in upper layer VI of the mouse primary somatosensory cortex (White and Keller, 1987).

identified corticothalamic projection cells; on the basis of certain physiological and morphological characteristics, Gabbott et al. (1987) tentatively identified the neurons they examined as corticotectal cells; and the projection sites of the layer VI neurons examined by McGuire et al. (1984) are unknown, but given the fact that the somata of corticotectal cells do not occur in layer VI (Figure 2.3) and that as many as two-thirds of the neurons in layer VI of the cat striate cortex apparently project to regions of the brain other than to the thalamus (see discussion in White and Keller, 1987), it is possible that the neurons examined by McGuire et al. (1984) project neither to the tectum nor to the thalamus. Other possibilities for the type of neurons examined by McGuire et al. include pyramids that project to the claustrum or to other areas of the cortex (see Figure 2.3). Neurons that project to different sites have different input relationships (see White, 1986)—that such cells might have different local output relationships appears reasonable. Accordingly, if the populations of axon terminals examined in each of the studies cited in Figure 3.3 belong to pyramidal cells that project to three different sites, it might be expected that their axon terminals would participate in three different and contrasting types of synaptic relationships.

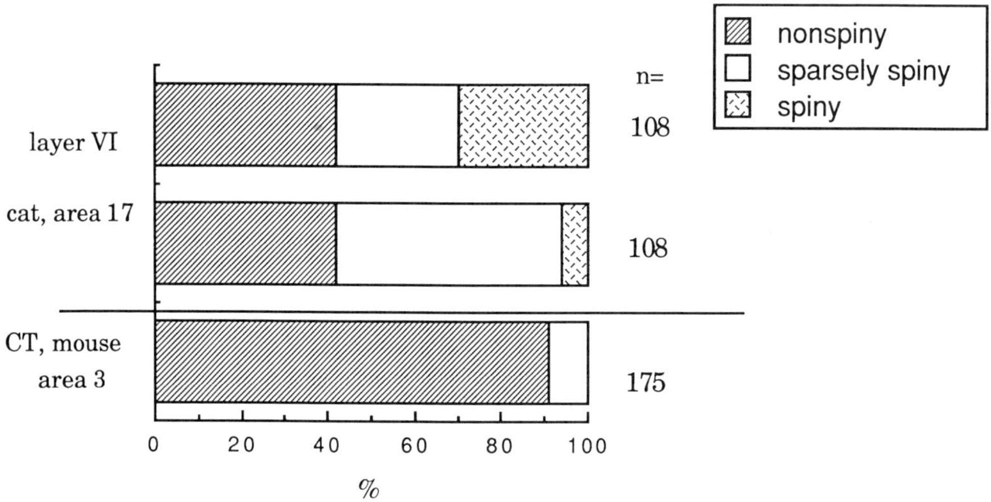

Figure 3.4. *Graph showing the percentages of axodendritic (shaft) synapses formed onto dendrites belonging to nonspiny, sparsely spiny, and spiny neurons by the axon collaterals of deep pyramidal cells. Sparsely spiny dendrites are those having <1 spine per 5 μm.*

The two bars above the horizontal line are from McGuire et al. (1984); the bar beneath the line is from White and Keller (1987). For the top bar, sparsely spiny dendrites having a low synaptic density (e.g., <1 synapse/2 microns) were considered to belong to spiny neurons. This interpretation is consistent with patterns of spines and synapes displayed by dendrites of spiny neurons, such as pyramids and spiny stellate cells (e.g., White and Rock, 1980; Hersch and White, 1981c).

n = numbers of synapses.

Elements postsynaptic to the axon collaterals of deep pyramidal neurons were identified by the same approaches used to identify elements postsynaptic to the axon collaterals of superficial pyramidal neurons. Thus, as before, reliance was placed mainly on the three-dimensional morphology of the postsynaptic elements and on the patterns of their synaptic connections as assessed in reconstructions from serial thin sections.

Axon collaterals of deep pyramidal neurons synapse onto nonspiny dendrites, and in two instances these dendrites were traced to their parent cell bodies. On the basis of the shapes of these cell bodies, one postsynaptic neuron was identified as a nonspiny multipolar neuron (White and Keller, 1987), and the other as a nonspiny bipolar neuron (McGuire et al., 1984). Synapses with nonspiny dendrites are presumed to indicate a circuit involving deep pyramidal neurons and any of a number of nonspiny or sparsely spiny nonpyramidal neuronal types (Figure 3.5) including the two types mentioned above, but also others such as basket cells, chandelier cells, double bouquet cells, etc.

Each of the three studies also reports synapses from pyramidal cell axon collaterals onto spines. Unequivocal evidence that some of these spines belong to other deep pyramidal cells has been provided for the corticothalamic cells (White and Hersch, 1982; White and Keller, 1987) and for both the neurons examined by Gabbott et al. (1987). That some of the spines belong to spiny or sparsely spiny nonpyramidal cells cannot be excluded.

Both studies of the cat visual cortex describe synapses from the axon collaterals of deep pyramidal cells onto the shafts of spiny dendrites. The cell type of origin of these dendrites is uncertain, but likely candidates include both spiny stellate and pyramidal cells. Shafts of spiny stellate cells are known to receive asymmetrical synapses (Benshalom and White, 1986), and despite the common perception that dendritic shafts of pyramidal neurons do not receive this synaptic type, it has been observed that approximately 2% of the synapses onto the dendritic shafts of layer V and VI pyramidal neurons are asymmetrical (Hersch and White, 1981b).

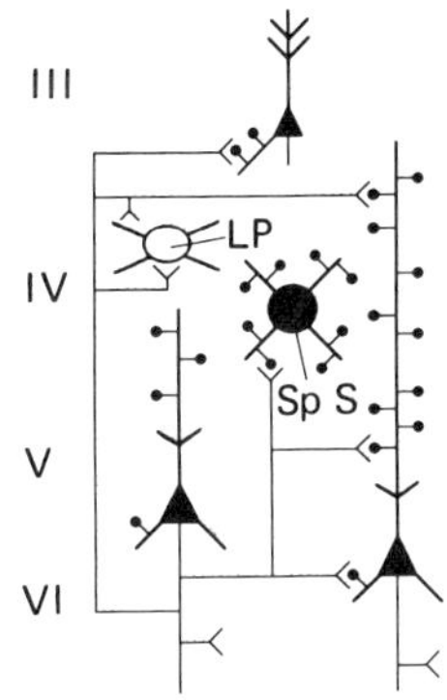

Figure 3.5. *Schematic illustration of the local projection of deep pyramidal neurons to other deep pyramids, to superficial pyramids, to spiny stellate cells (SpS), and to local plexus neurons (LP). Because they are relatively uncommon, and to simplify the diagram, synapses onto dendritic shafts of the spiny neurons have been omitted.*

For reasons having to do with the laminar distribution of dendrites belonging to pyramidal neurons versus spiny stellate cells (see Chapters 1 and 2), it is reasonable to conclude that the local output of deep pyramidal neurons to spiny dendrites, in layers V and VI, consists mainly of circuits connecting deep pyramids with other deep pyramids, and in layer IV, consists, in addition, of circuits connecting deep pyramids with spiny stellate cells and, perhaps by way of their basal dendrites, with superficial pyramidal cells (Figure 3.5). Above layer IV, the axon collaterals of deep pyramidal neurons presumably synapse with spines and shafts of spiny dendrites belonging to both superficial and deep pyramidal neurons. In all instances, the synapses made by the local axon collaterals of deep pyramidal cells are of the asymmetrical type, and so the circuits that have been described can be considered as representing the anatomical substrates of feedforward excitation (Ferster and Lindström, 1985).

Summary

The axon collaterals of pyramidal neurons display a variety of synaptic output relationships: In certain instances synapses are made mainly with spines or mainly with dendritic shafts; in other instances these postsynaptic elements are contacted in about equal proportions. At present, no relationship can be discerned between a particular pattern of synaptic output and the laminar location either of the local axonal ramification or of the parent cell body. However, a relationship may exist between the synaptic output patterns of local axon collaterals and the distant projection site of the main axon.

The axon collaterals of pyramidal neurons form asymmetrical synapses onto other pyramidal neurons and onto various types of nonpyramidal neurons including those that are GABAergic (Keller and White, 1989). These synaptic connections provide the anatomical substrate for the feedforward excitation of cortical pyramids and for feedforward and feedback inhibition via connections with GABAergic neurons.

Intrinsic Connections of Nonpyramidal Neurons

The kinds of synapses involving the local axonal ramifications of nonpyramidal neurons have been presented in Chapter 2 in the separate sections that describe the morphological aspects of various nonpyramidal neuronal types. A summary of these connections is presented in Table 3.1.

Data on the frequency with which the axons of nonpyramidal neurons synapse with different, identified postsynaptic elements have been provided in several of the studies cited in Table 3.1. Somogyi et al. (1983a) report that, of 177 elements postsynaptic to the local axon collaterals of two basket cells in cat visual cortex, 33% synapsed with pyramidal cell bodies, 24% with their dendritic shafts, and 20% with spines of unknown origin; the remainder were presynaptic to various elements including dendrites and cell bodies of nonspiny nonpyramidal neurons and the initial axonal segment of a pyramidal neuron. Similar findings are reported by Freund et al. (1986) for basket cells shown to be immunoreactive for cholecystokinin. Subsequently, Kisvarday et al. (1987) report that, of 199 elements postsynaptic to the local axon collaterals of a single basket cell in cat visual cortex, only about 20% synapsed with pyramidal cell bodies, whereas the values for dendritic shafts and spines, interpreted as mainly of pyramidal cell origin, were about 38% and 41%, respectively.

Discrepancies in the proportion of axosomatic synapses formed by basket cell axons may indicate that a range of values characterizes the specific local output relationships of cortical basket cells. Alternatively, it may be that basket cells in superficial cortical layers, that is, those examined by Somogyi et al. (1983a) and Freund et al. (1986), display synaptic output relationships that differ from those of basket cells in the deeper layers of cortex, examined by Kisvarday et al. (1987). Reports on the local output relationships of axons belonging to six "clutch" cells (a local plexus neuron considered a subtype of basket cell) show that, of 480 synapses, 35% to 58% were onto dendritic shafts, 20% to 46% were onto spines with 10% to 30% onto cell bodies and, rarely, onto initial axon segments (Kisvarday et al., 1985, 1986b). Some of these synapses were made with identified spiny stellate cells, the remainder were made with a variety of neurons presumed to include pyramidal,

Table 3.1. *Local synapses of cortical nonpyramidal neurons.*

Neuronal type	Species	Area	Synaptic type	Postsynaptic elements	References
Chandelier	Rat	VI	Sym	Axon initial segment of pyramidal cells[a]	Somogyi, 1979; Peters et al., 1982; Somogyi et al., 1982
	Rat	VII	Sym	Axon initial segment of pyramidal cells	Somogyi et al., 1982
	Cat	VI	Sym	Axon initial segment of pyramidal cells[a]	Fairén and Valverde, 1980; Freund et al., 1983
	Cat	VII	Sym	Axon initial segment of pyramidal cells	Somogyi et al., 1982
	Cat	MI	Sym	Axon initial segment of pyramidal cells	Somogyi et al., 1982
	Monkey	MI	Sym	Axon initial segment of pyramidal cells	DeFelipe et al., 1985
	Monkey	VI	Sym	Axon initial segment of pyramidal cells	Somogyi et al., 1982
	Monkey	VII	Sym	Axon initial segment of pyramidal cells	Somogyi et al., 1982
	Human	Temporal	Sym	Axon initial segment of pyramidal cells	Kisvarday et al., 1986a
Basket	Cat	VI	Sym	Cell bodies, dendrites, spines axon initial segment of pyramidal neurons; cell body, dendrites nonspiny nonpyramidal neurons; spines of unidentified origin	DeFelipe and Fairén, 1982; Freund et al., 1986; Kisvarday et al., 1987; Somogyi et al., 1983a; Somogyi and Soltész, 1986
	Monkey	SI	Sym	Cell bodies and proximal dendrites of pyramidal neurons; cell bodies of nonspiny nonpyramidal neurons	DeFelipe et al., 1986b
Nonspiny and sparsely spiny local plexus[b] (multipolar)	Rat	VI	Sym	Cell bodies, proximal dendrites and axon initial segment of pyramidal neurons; cell bodies, dendrites of nonspiny nonpyramidal neurons; spines of unknown origin	Peters and Fairén, 1978; Peters and Proskauer, 1980; Ribak, 1978
Clutch[c] cell	Cat	VI	Sym	Cell bodies of pyramidal and presumably of spiny stellate and of nonspiny nonpyramidal cells; dendrites, axon initial segments, and spines of unknown origin	Kisvarday et al., 1985
	Monkey	VI	Sym	Spiny stellate cell bodies and dendrites; spines of unknown origin	Kisvarday et al., 1986b
Double bouquet with mainly descending axons	Cat	VI	Sym	Cell body and dendrites of pyramidal and spiny nonpyramidal cells; cell bodies and dendrites of nonspiny nonpyramidal neurons; spines of unknown origin	Freund et al., 1986; Somogyi and Cowey, 1981

Table 3.1. *Continued.*

Neuronal type	Species	Area	Synaptic type	Postsynaptic elements	References
	Monkey	VI	Sym	Cell bodies and dendrites of nonspiny nonpyramidal neurons and spines of unknown origin	Somogyi and Cowey, 1981
Double bouquet with ascending and descending axons	Cat	VI	Asym	Spines of unknown origin	Fairén et al., 1984
Bipolar	Rat	VI	Asym	Spines and shafts of pyramidal neurons; dendrites and cell bodies of nonspiny nonpyramidal neurons	Peters and Kimerer, 1981; Peters and Harriman, 1988
	Rat	VI	Sym	Dendritic shafts of nonpyramidal neurons	Peters and Harriman, 1988
Spiny stellate	Mouse	SI	Asym	Spines of unknown origin shafts of spiny and nonspiny dendrites	White and Rock, 1980;
	Cat	VI			LeVay, 1973;
	Monkey	VI			Levay, 1973; Saint Marie and Peters, 1985

[a]A single spine in rat VI and a single dendrite in cat VI received synapses from chandelier cell axon terminals (Somogyi et al., 1982).
[b]Includes putative basket cells.
[c]Term applied by Kisvarday et al. (1985) to a small type of basket cell whose axon is restricted to layer IV; for additional information regarding classification of basket cell types, see Somogyi and Soltész, 1986.

spiny, and nonspiny nonpyramidal types. In none of these instances did basket cell axons form more than a third of their synapses with the cell bodies of pyramidal neurons. This finding is at variance with the general perception that these neurons synapse primarily with the cell bodies of cortical pyramids.

Double bouquet cells with descending axons are presynaptic mainly to spines (53.8%) and to dendrites (39.3%) and less frequently to cell bodies (6.9%) (Freund et al., 1986). Of a total of 53 synapses made by spiny stellate cells in rat visual cortex, 66% were onto spines; the remainder were onto dendritic shafts of both the spiny and nonspiny types (Saint Marie and Peters, 1985). Finally, it may be supposed on the basis of light microscopic results that single chandelier cells in the visual cortex of the cat (Freund et al., 1983) and in the motor cortex of the monkey (DeFelipe et al., 1985) each synapse with initial axon segments belonging to from 300 to 400 pyramidal neurons, although the possibility that some of these terminals synapse onto nonpyramidal neurons cannot be excluded (cf. Lund, 1987).

DeCarlos et al. (1987) noted that, in addition to forming vertical rows of boutons that appear to contact initial axon segments, chandelier cell axons may form curvilinear arrays of terminals that resemble the "basket type endings" known to form synapses onto cell bodies. These curvilinear arrays may synapse onto the somata of pyramidal neurons, but a nonpyramidal cell target cannot be excluded.

Summary

Recent findings have dispelled certain myths regarding the local output relationships of nonpyramidal neurons: Axons of basket cells form but a third or less of their synapses with the somata of pyramidal neurons, and the axons of chandelier cells, which had been thought to be quite specific for the initial axon segments of pyramidal neurons, exhibit morphologies that suggest that they too may form synapses with disparate elements. Both of these findings lend support to the contention that axons of nonpyramidal neurons are relatively nonselective with regard to the type of postsynaptic elements with which they synapse: In general, axons of nonpyramidal neurons make synapses onto spines, dendritic shafts, and cell bodies of both pyramidal and nonpyramidal neurons.

Relatively little is known of the specific identity of elements postsynaptic to the intrinsic axonal projections of nonpyramidal neurons. A corollary of our position that information processing within the cerebral cortex is mediated primarily by anatomically stable circuits is that these circuits are composed of specific types of synaptic connections. Presumably, the specificity of synaptic connections involving the axons of nonpyramidal neurons will become increasingly apparent as additional information is obtained on both the qualitative and quantitative aspects of their synaptology.

Efforts to elucidate the intrinsic circuitry of the cerebral cortex have borne fruit only recently and without doubt many local cortical pathways, and the specifics of their interneuronal connections, have yet to be discovered. The main reason for this has to do with limitations inherent in the approaches used to identify pre- and postsynaptic elements involved in intrinsic synaptic connections, that is, synaptic connections between two neurons whose somata occur in proximity within the same region of the cortex. Most data on intrinsic circuits have been obtained from preparations in which the presynaptic element has been labeled by the Golgi method (e.g., LeVay, 1973) or by the intracellular injection (e.g., McGuire et al., 1984) or retrograde transport (e.g. White et al., 1980) of horseradish peroxidase.

The identification of postsynaptic elements has relied primarily on the making of reconstructions from long series of thin sections through unlabeled profiles (e.g., Keller and White, 1987; Somogyi, 1978; White and Keller, 1987). The main problem with using this approach to identify unlabeled profiles is that even extensive reconstructions of unlabeled elements often lack critical information needed to identify the neuronal type of the reconstructed process. Frequently, what is needed to identify an unlabeled process is information on the shape and synaptic patterns of its parent cell body, but because of the distances involved relative to the thinness of the sections required by electron microscopy, it is usually not possible to trace neuronal elements back to their origins. The ideal situation is to use preparations in which both the pre- and postsynaptic neurons are labeled (e.g., Keller and White, 1989; Peters and Proskauer, 1980; Somogyi et al., 1979). However, in order to avoid confusion, the labeled pair of neurons must exist in relative isolation from other labeled elements, a situation that is frequently unattainable with available methodologies. Thus, for a variety of reasons, the elucidation of intrinsic cortical connections persists as one of the greatest challenges facing workers in synaptic organization.

Synapses of Extrinsic Afferents With Cortical Neurons

The situation regarding synapses between extrinsic neuronal elements and cortical neurons is much more favorable primarily because by comparison with intrinsic pathways, extrinsic pathways and sources are far more amenable to experimental manipulation. This feature of the extrinsic pathways is a direct consequence of the spatial separation between the sources of extrinsic input and their sites of termination within the cerebral cortex. The strict separation between the sources and terminations of extrinsic cortical pathways made it possible to obtain information regarding the major input pathways to the cortex by relatively crude methods beginning more than a century ago, and it is safe to say that by now all the major input pathways to the cerebral cortex have been identified. For instance, as early as 1870 (Gudden), destruction of the cortex with consequent retrograde cell degeneration in the thalamus was used successfully to provide experimental evidence of a functional link between the thalamus and cortex. Subsequently, investigations using a wide variety of more sensitive neuronal labeling methods have identified the specific sources and the precise laminae of termination of afferents to the cortex from both cortical and subcortical regions of the brain (e.g., see Innocenti, 1986; White, 1979).

The advantage conferred by having the source and termination of extrinsic afferent pathways separated by relatively large distances also has had a salutory effect on attempts to identify the specific types of cortical neurons that receive synapses from extrinsic sources. Thus, in contrast to the situation regarding intrinsic cortical connections, labeling of the presynaptic elements in an extrinsic circuit can be effected readily in thin sections by relatively simple, reliable methods that do not impinge on the area of cortex under study, a factor that helps ensure against the nonselective labeling of neuronal elements. These methods include approaches whereby the parent cell body of the presynaptic element is lesioned (cf. lesion-induced degneration [Colonnier, 1964]) or filled with a labeling substance that is then transported anterogradely down the axon to the presynaptic terminals (cf. anterograde transport methods involving radioactive amino acids [Hendrickson, 1972, 1975], lectins [Cipolloni et al., 1985], or intracellularly injected HRP [Landry and Deschênes, 1981]). Results to date indicate that, in every system examined, axon terminals belonging to extrinsic afferents to the cerebral cortex invariably form only asymmetrical, presumed excitatory synapses (e.g., Colonnier, 1981; for a possible exception, see Einstein et al., 1987).

At present, the extrinsic pathways whose synaptic connections within the cerebral cortex have received the most attention are those originating in the specific nuclei of the thalamus and terminating within the primary sensory areas of the neocortex. The emphasis on these thalamocortical pathways no doubt has its roots in the observation that cytoarchitecturally distinct zones of the cerebral cortex that are involved in different functional activities are each related to a particular region or regions of the thalamus (see White, 1979). The relationship of thalamic nuclei with cytoarchitectonic regions is perhaps best illustrated by the thalamic projection to the primary sensory areas of the neocortex: Each of these areas receives a particularly intense, topographically organized projection from a specific thalamic nucleus.

The realization that thalamocortical input to the primary areas of the cerebral cortex plays a significant role in the organization of neurons in these regions into vertically oriented functional columns or bands (e.g., see Mountcastle, 1978, 1979) has motivated a variety of attempts to identify the sequences of synaptic connections through which thalamic input enters the columns and is then processed within them. The remainder of this section will be concerned primarily with circuitry involving thalamocortical afferents, both because so much is known about this aspect of cortical circuitry and because available information indicates that many of the features of the circuitry involving thalamocortical afferents are applicable also to circuits involving other extrinsic afferent systems to the cerebral cortex.

Initially, efforts to identify neuronal types that receive synapses from thalamocortical afferents employed light microscopic analyses of Golgi-impregnated neurons in tissue taken from the cortices of sensory deprived animals or from animals that had sustained damage at some point along an ascending sensory pathway. Changes in the morphologies of impregnated spiny stellate and pyramidal neurons were interpreted to mean that both of these cell types form thalamocortical synapses (see the review by White, 1979, p. 288).

However, the reliability of these interpretations was uncertain because the light microscopic methods employed lacked the resolution to visualize synaptic contacts. For instance, Globus and Scheibel (1967) observed that eye enucleation in newborn rabbits causes a decrease in the numbers of spines along the middle portions of the apical dendrites of layer III pyramidal cells in primary visual cortex. The midportions of these apical dendrites occur in laminae known to receive little or no direct thalamocortical input, and so it seems that transneuronal effects, rather than direct deafferentation, were the bases for the decrease in spine numbers. As seen earlier (see "Summary," Chapter 2), the coexistence of neuronal processes within close proximity is necessary but insufficient evidence on which to presume the existence of a synaptic contact, and so observations made with the light microscope at best only imply the existence of a synaptic contact between specific neuronal elements. Unequivocal morphological evidence that the synapse is actually present must come from electron microscopy.

The application of electron microscopy to the task of identifying neuronal types that directly receive thalamocortical input involved, at first, the examination of thin sections through tissue containing thalamocortical axon terminals that had been labeled by lesion-induced degeneration. It soon became evident (see White, 1979, p. 289) that both spiny and nonspiny neurons receive thalamocortical synapses. However, the methods employed in these studies did not allow the three-dimensional forms of neurons to be determined, and so, for the most part, the specific identity of neurons postsynaptic to thalamic afferents could not be ascertained.

One approach to overcoming this difficulty involves reconstructing the postsynaptic neurons from extensive series of thin sections. Using this method, Davis and Sterling (1979) identified thalamocortical synapses onto dendrites belonging to pyramidal cells and onto several types of nonpyramidal cells of the bitufted and multipolar varieties. Another approach to identifying neurons that receive thalamocortical synapses involves the combination of different methods to label both the pre- and postsynaptic elements of the thalamocortical synapse. Most thalamocortical synapses are made with dendritic spines (e.g., in primary sensory [Peters and Feldman, 1976] and motor [Strick and Sterling, 1974] areas of mammals, in the reptile cortex [Smith et al., 1980]); and because it is usually not possible to trace *un*labeled spines back to their parent dendrites, the labeling of postsynaptic elements has proved essential to identifying thalamocortical synapses with spiny neuronal types.

The two methods used most commonly to label neurons postsynaptic to thalamocortical afferents are the Golgi-EM technique combined with gold toning (Fairén et al., 1977) and the retrograde transport of horseradish peroxidase (HRP) (e.g., White, Hersch, and Rock, 1980). The Golgi-EM method has the advantage that it may be applied to label both pyramidal and nonpyramidal neurons, but suffers from the disadvantage that there is relatively little control over the types of neurons that become impregnated. In addition, lesioning the thalamus results, on the average, in the appearance of fewer impregnated neurons in the related region of the cortex (personal observations); what effect thalamic lesions might have on the types of neurons that are impregnated is unknown.

The retrograde HRP method is employed primarily to label pyramidal cells whose axons project to sites that can be injected with the enzyme, at some distance from the region of cortex being examined. However, the method has been applied also to label both projection and nonprojection neurons in the vicinity of an injection site within the cortex (e.g., Vanegas et al., 1977). HRP is also used as an intracellular marker in electrophysiological studies (e.g., see Parnavelas, 1984), but as yet this approach has not been combined with methods for labeling thalamocortical afferents. Findings of electron microscopic studies in which the labeling of thalamocortical afferents by lesion-induced degeneration or by anterograde filling with HRP has been combined with either the Golgi-EM or the retrograde HRP methods for labeling postsynaptic neurons indicate that a wide variety of neuronal types forms synapses with thalamocortical afferents.

In most instances, no information was available on the branching patterns of the axons of nonpyramidal neurons listed in Table 3.2. Because of this, nonpyramidal cells were classified using terms that describe primarily the shape of their cell bodies and the distribution of their dendrites. For this reason, neurons classified as smooth or sparsely spiny multipolar probably include cells that might have been labeled as chandelier, basket, or local plexus cells had information been available on the distribution of their axonal branches.

All the data reported in Table 3.2 involve synapses between labeled thalamocortical afferents that originate in specific thalamic nuclei and the spines, dendrites, and/or cell bodies of identified neurons in the primary sensory areas of the neocortex. These findings, which include investigations of thalamocortical connectivity in different cortical areas in several species, indicate that every neuron having a dendrite in layer IV of the primary sensory areas of the neocortex forms some proportion

Table 3.2. *Neurons receiving thalamocortical synapses in layer IV of primary sensory areas of the cortex.*

Cell type	Layer of cell body	Species	Area	References
Pyramid	III	Mouse	SmI	White, 1978; White and Hersch, 1981
	III	Rat	VI	Peters et al., 1979
	III	Cat	VI	Hornung and Garey, 1980, 1981
	III	Cat	VII	Freund et al., 1985
	III	Monkey	SI	Hendry and Jones, 1980, 1983
	III/IV	Mouse	SmI	White, 1978; White and Hersch, 1981
	III/IV	Cat	VI	Davis and Sterling, 1979
	IV	Mouse	SmI	White and Hersch, 1981
	IV	Rat	VI	Somogyi, 1978
	V	Mouse	SmI	White, 1978; Hersch and White, 1981a, b, c, 1982; White and Hersch, 1982
	V	Rat	VI	Peters et al., 1977, 1979; Schober et al., 1983a
	V	Cat	VI	Hornung and Garey, 1980, 1981
	VI	Mouse	SmI	Hersch and White, 1981a, b; White and Hersch, 1982; White and Keller, 1987
Nonpyramid				
Smooth multipolar	IV	Mouse	SmI	White, 1978; White et al., 1984; Keller and White, 1987
	IV	Cat	VI	Hornung and Garey, 1980, 1981; Einstein et al., 1987; Freund et al., 1985
	V	Mouse	SmI	White and Rock, 1981
Sparsely spinous multipolar	IV	Rat	VI	Peters et al., 1979
	IV	Cat	VI	Hornung and Garey, 1980, 1981
Spiny bitufted	IV	Cat	VI	Hornung and Garey, 1980, 1981
	V	Mouse	SmI	Hersch and White, 1981a
Sparsely spinous bitufted	III	Rat	VI	Peters et al., 1979
Nonspiny bitufted	IV	mouse	SmI	White and Rock, 1981
	IV	Cat	VI	Davis and Sterling, 1979
Nonspiny bipolar	IV/V	Mouse	SmI	White, 1978
	IV	Rat	VI	Peters et al., 1976, 1979; Schober et al., 1983b
Spiny stellate	IV	Mouse	SmI	White, 1978; White and Rock, 1979, 1980; Benshalom and White, 1986
	IV	Cat	VI	Hornung and Garey, 1980, 1981

of its synapses with thalamocortical afferents (cf. White, 1987). Postsynaptic neurons include both pyramidal and nonpyramidal neurons whose somata occur not only in layer IV, but also in layers of cortex above and below layer IV. Evidence from electrophysiological studies confirming that both pyramidal and nonpyramidal neurons receive monosynaptic input from the thalamus is provided in Part II (Table 5.1). Thus, ample findings exist to refute the suggestion that spiny stellate cells are the sole or even the main recipients of thalamic input (cf. Eccles, 1984). This relatively new view of thalamocortical synaptic relations provides a direct and important challenge to the concept that thalamic input to the cortex is processed by hierarchically organized chains of neurons (see Part III).

Comparatively little information is available regarding synaptic connections involving other afferent systems that project to the primary sensory areas or to other regions of the cerebral cortex, but in general, what is known is consistent with the notion that afferents to the cortex form asymmetrical, excitatory synapses with both pyramidal and nonpyramidal cells. For example, terminals of thalamocortical afferents to the primary motor cortex in the cat synapse at asymmetrical junctions with spines belonging to pyramidal neurons and with nonspiny dendrites belonging to nonpyramidal cells (Ichikawa et al., 1985; Strick and Sterling, 1974; and cf. findings in monkey, Sloper, 1973). Kosar et al. (1985) recorded monosynaptic responses from neurons in cat motor cortex following stimulation of the thalamus. Injecting label into the postsynaptic neurons demonstrated that thalamocortical afferents synapse directly onto both pyramidal and nonpyramidal neurons.

Afferents to the primary visual cortex of the cat from the nucleus centralis, a nonspecific thalamic nucleus (Cunningham and LeVay, 1986), and from the claustrum (LeVay, 1986) also synapse at asymmetrical junctions, mainly with spines, but also with dendritic shafts. Callosal afferents to the primary motor (Sloper, 1973; Sloper and Powell, 1979) and primary visual (Fisken et al., 1975) cortex of the monkey, to the primary and secondary somatosensory cortex of the cat (Jones and Powell, 1970), and to the secondary visual (Lund and Lund, 1970) and primary auditory cortex of the rat (Cipolloni and Peters, 1983) form nearly all their synapses with spines thought to belong to pyramidal cells, but a few are formed with nonspiny dendrites and, rarely, with cell bodies that have been interpreted to belong to nonpyramidal cells. Afferents connecting ipsilateral areas of the cortex in the mouse (Porter and White, 1984, 1986), cat (Ichikawa et al., 1985; Jones and Powell, 1970), and monkey (Fisken et al., 1975; Sloper, 1973; Sloper and Powell, 1979) also synapse mainly with spines, most likely of pyramidal neurons and to a lesser extent with nonspiny dendrites presumed to belong to nonpyramidal neurons.

With a few exceptions (e.g., Cipolloni and Peters, 1983; Ichikawa et al., 1985; Kosar et al., 1985; Porter and White, 1984, 1986), postsynaptic elements were not labeled in these studies, and so the identification of neuronal types has been based, for the most part, on indirect evidence. For instance, the assumption that unlabeled spines belong to pyramidal cells derives mainly from evidence showing that pyramidal cells are the most common type of spine-bearing neuron in the neocortex (e.g., Peters et al., 1985). Thus, at least for layers of cortex outside of layer IV, which is known to contain large numbers of dendrites belonging to spiny stellate cells, the presumption has been that spines of uncertain origin are likely to belong to pyramidal cells.

The identification of unlabeled postsynaptic dendrites as of nonpyramidal origin is somewhat more problematical, because it is based mainly on the assumption that only dendrites belonging to certain types of nonpyramidal cells do not have spines, but that dendrites of all other neuronal types do. However, it has been clear for some time that the proximal dendrites of pyramidal cells may have very few spines (e.g., Cajal, 1909–1911), and it has been shown also that middle and distal portions of pyramidal cell dendrites may in some instances lack spines (Hersch and White, 1981b). For this reason, some nonspiny dendritic segments may in fact be dendrites of pyramidal neurons.

Another feature used to differentiate dendrites of pyramidal neurons from those belonging to nonpyramidal neurons is related to the distribution of asymmetrical synapses on their surfaces: dendrites of pyramidal neurons usually receive asymmetrical synapses only on their spines; in contast, nonspiny, nonpyramidal neurons typically receive large numbers of asymmetrical synapses directly onto their shafts (e.g., Feldman, 1984; White et al., 1984). However, results of studies using the Golgi-EM method combined with reconstructions from serial thin sections show that some parts of pyramidal cell dendritic shafts are postsynaptic at relatively high frequency to asymmetrical synapses (e.g., Hersch and White, 1981b). Because neither the numbers of spines nor the numbers of asymmetrical shaft synapses are foolproof criteria for the identification of dendritic segments, it is likely that in some instances, unlabeled, nonspiny segments of pyramidal cell dendrites have been mistakenly identified as of nonpyramidal origin.

Summary

Evidence based on the unequivocal identification of synapses between labeled pre- and postsynaptic elements shows that thalamocortical afferents to the primary sensory areas of the neocortex synapse with many types of pyramidal and nonpyramidal neurons. In contrast, some degree of uncertainty surrounds the identification of elements postsynaptic to other afferent pathways that project to the primary sensory areas or to other areas of the neocortex. Nevertheless, sufficient evidence exists to support the broad contention that axon terminals of all types of extrinsic afferents form asymmetrical, presumed excitatory synapses with both pyramidal and nonpyramidal cells.

Quantitative Aspects of Circuity Involving Extrinsic Cortical Afferents

Regardless of the identity of the neurons contacted, it is clear beyond any doubt that the main synaptic target for all extrinsic cortical afferents is spines. This observation has been confirmed for a variety of extrinsic afferent pathways in several species by studies that have provided data on the numbers and proportions of synapses formed by extrinsic cortical afferents onto spines and dendritic shafts (e.g., see Colonnier, 1981, Table I). Examination of these data shows that despite the apparent similarities in the organization of afferent input to the cortex—that is, at all termination sites of extrinsic pathways, axospinous synapses predominate over axodendritic ones—there are some significant differences in the proportions of axospinous versus axodendritic synapses formed by the different pathways. For instance, afferents belonging to ipsilateral pathways linking sensory and motor areas of monkey cortex form roughly 80% of their synapses with spines (Sloper and Powell, 1979), whereas the percentage of axospinous synapses formed by callosal afferents to these areas is approximately 97% (Sloper and Powell, 1979). The percentage of axospinous synapses formed by thalamocortical afferents to the sensory and motor regions of the monkey cortex is, at about 90%, approximately midway between these values, and in general, thalamocortical afferents form a larger proportion of their synapses with presumed nonpyramidal cell bodies than do afferents belonging to either ipsilateral or callosal pathways (Colonnier, 1981; Sloper and Powell, 1979).

As mentioned earlier in connection with intrinsic synapses, comparisons of the percentages of axospinous and axodendritic synapses at best may be uninformative and at worst misleading in the absence of precise information regarding the composition of the neuropil within the target region. For this reason it may turn out to be irrelevant to compare, as has been done here, synaptic patterns of thalamocortical afferents with those of corticocortical and callosal afferents because these elements terminate within different cortical layers whose homogeneity of composition has yet to be established.

Similar considerations would apply to comparisons of the synaptic output relationships of a single pathway in different regions. For example, claustral afferents to the primary visual cortex of the cat (LeVay, 1986) form, in layers I and VI, respectively, 96% and 88% of their synapses with spines, and the remainder with dendritic shafts; claustral afferents to layer IV are presynaptic in about equal proportions to spines and dendritic shafts (LeVay, 1986). These findings indicate that axon terminals belonging to a single, afferent pathway can terminate with different frequencies onto spines and dendrites within separate layers of the cortex; however, without knowing something of the composition of the neuropil in the target regions, it is impossible to determine whether the findings merely reflect the relative concentrations of axospinous and axodendritic synapses in the neuropil or whether some form of selectivity is involved in generating the observed proportions of different synaptic types. Of course, when target regions coincide, this caveat does not apply or applies only in part: Findings showing that claustral afferents to layer IV of the cat visual cortex are presynaptic in about equal proportions to spines and dendritic shafts (LeVay, 1986), whereas thalamocortical afferents to the same area form roughly 90% of their synapses with spines (Garey and Powell, 1971; LeVay, 1986; Winfield and Powell, 1983) indicate that at least one of these afferent types distributes its synapses selectively, that is, without regard for the concentration in the neuropil of axospinous and axodendritic synapses.

These findings imply that, despite the propensity of external afferents to synapse with many different types of cortical neurons, these afferents are likely to be highly selective with regard to the kinds of targets with which they synapse (i.e., spines vs. dendritic shafts). That this is indeed the case has been made abundantly clear by the findings of a series of studies aimed at elucidating the numbers and distribution of synapses between thalamocortical afferents and identified neurons in the mouse primary somatosensory (SmI) cortex.

Most quantitative information currently available on the synaptic connectivity of thalamocortical afferents is derived from the mouse SmI cortex, in large part because in this region of the cortex, it has proved possible to use relatively simple methods to identify all thalamocortical axon terminals. The usual approach has been to lesion the ventrobasal thalamus, which contains the specific nucleus that projects to the SmI cortex, and then to have the animal survive several days until the thalamocortical axon terminals in ipsilateral SmI cortex have degenerated sufficiently to be recognized. Because thalamocortical axon terminals[1] to the mouse SmI cortex degenerate within a narrow time course, they can all be identified at any single, appropriate postlesion survival time

[1]The terminals referred to here, and elsewhere in the work of White and his colleagues, are from specific thalamocortical afferents that ramify mainly in layer IV, within the hollows of barrels in the posteromedial barrel subfield as defined by Woolsey and Van der Loos (1970). Evidence that the specific thalamic projection terminates within barrel hollows, whereas the nonspecific projection from the posterior thalamic nucleus projects to the septa between barrels, has been provided by Lin et al. (1987) and cf. Lorente de Nó (1938).

(e.g., White, 1978). Results of this approach typically show that approximately 20% of the asymmetrical synapses in layer IV of the mouse SmI cortex are formed by thalamocortical axon terminals (e.g., White, 1981, 1986), a value confirmed by a study in which the anterograde transport of the lectin *Phaseolus vulgaris*-leucoagglutinin was used to label thalamocortical axon terminals of the same projection system (Keller et al., 1985).

The use of lesion-induced degeneration for quantitative studies has thus far been limited to the mouse, because in other systems thalamocortical axon terminals degenerate over a varied time course such that some terminals have been phagocytosed before others have begun to degenerate (e.g., Jones and Powell, 1970; Tigges and Tigges, 1979). This means that only a fraction of the total number of thalamocortical synapses can be identified at any single postlesion survival time. A number of interesting theories have been advanced as to why this is so, but a discussion of them is beyond the scope of this work.

Quantitative studies of the connections of extrinsic afferents in that mouse primary somatosensory cortex have included examinations of thalamocortical synapses with both pyramidal and nonpyramidal types of neurons. Pyramidal neurons have been identified by impregnation with the Golgi method (e.g., Hersch and White, 1981b) or by the retrograde transport of HRP (e.g., White and Hersch, 1981). Nonpyramidal neurons have been identified either by impregnation with the Golgi method (e.g., White and Rock, 1980), by their appearance in reconstructions made from series of thin sections that included their cell bodies as well as extensive portions of their dendritic trees (e.g., White and Rock, 1981), or by their content of GABA, which to date has been observed only within nonpyramidal neuronal types (Keller and White, 1987).

In general, the approach has been to make three-dimensional or graphic reconstructions of dendrites from preparations in which thalamocortical axon terminals have been labeled. Plotting different synaptic types onto the reconstructions in their appropriate positions makes it possible to determine the frequency and the spatial distribution of synapses made with the reconstructed dendrites.

One aspect of the synaptic connectivity of afferent pathways that can be examined by the quantitative approaches described above concerns the variability of synaptic inputs to different dendrites of a single neuron. Dendrites that cross several laminae of the cortex, for example, the apical dendrites of pyramidal neurons, are exposed to a variety of afferent types (see Chapter 1), and so it would not be unexpected for different dendrites of pyramidal neurons (e.g., apical vs. basal dendrites, the different side branches of an apical dendrite) to be involved in different synaptic relationships, depending on the laminae within which the dendrites occur. In contrast, many types of nonpyramidal neurons have all their dendrites within a single cortical lamina creating a condition whereby each dendrite is exposed to a similar set of inputs.

Under these conditions, it would be interesting to know whether different dendrites belonging to the same neuron exhibit different synaptic relationships. This question has been examined in several studies, and in every instance the results are consistent with the notion that different dendrites of a single neuron possess markedly similar synaptic relationships, provided the dendrites occur within a single cortical lamina. For example, each of four dendrites belonging to a nonspiny bitufted cell in layer IV of the mouse primary somatosensory cortex receives a little over 20% of its asymmetrical synapses from thalamocortical axon terminals. In the same region, two dendrites belonging to a nonspiny multipolar neuron receive just under 4% of their synapses from the same source (White and Rock, 1981). Additional findings confirm that the display of similar proportions of thalamocortical and other synapses by different dendrites of a single neuron is a general feature of nonpyramidal neurons whose dendrites occur within a single cortical layer (cf. thalamocortical and other synaptic relationships of different dendrites belonging to a spiny stellate cell [White and Rock, 1980] or to several nonspiny multipolar or bitufted neurons [Keller and White, 1987; White et al., 1984; White and Rock, 1979, 1981].

The similarity in the synaptic relationships of different dendrites of a single neuron is not limited to the proportions of synaptic types formed, but includes also the spatial relationships of the different types of synapses along the postsynaptic dendrites. This is illustrated graphically by Figure 3.6, which plots the distribution of synapses along two dendrites belonging to the nonspiny multipolar neuron examined by White and Rock (1981). Figure 3.6 was constructed by performing a five-bin boxcar integration for the total number of synapses, for symmetrical synapses, and for thalamocortical synapses within 3-μm bins beginning at the cell body. The distribution of asymmetrical synapses was omitted for the sake of clarity. Because asymmetrical synapses account for most of the synapses along the entire portions of both dendrites, the distribution of asymmetrical synapses roughly parallels the plot for "all synapses." The shapes of the curves for the different categories of synapses are similar for the two dendrites, indicating that the distribution of synaptic types is similar for dendrites that belong to the same neuron.

The absolute numbers of synapses in each category are lower for the dendrite shown in Figure 3.6b than for the dendrite shown in Figure 3.6a. Presumably, the numbers of synapses formed by a dendrite depend not only on the affinity of the pre- and postsynaptic elements for each

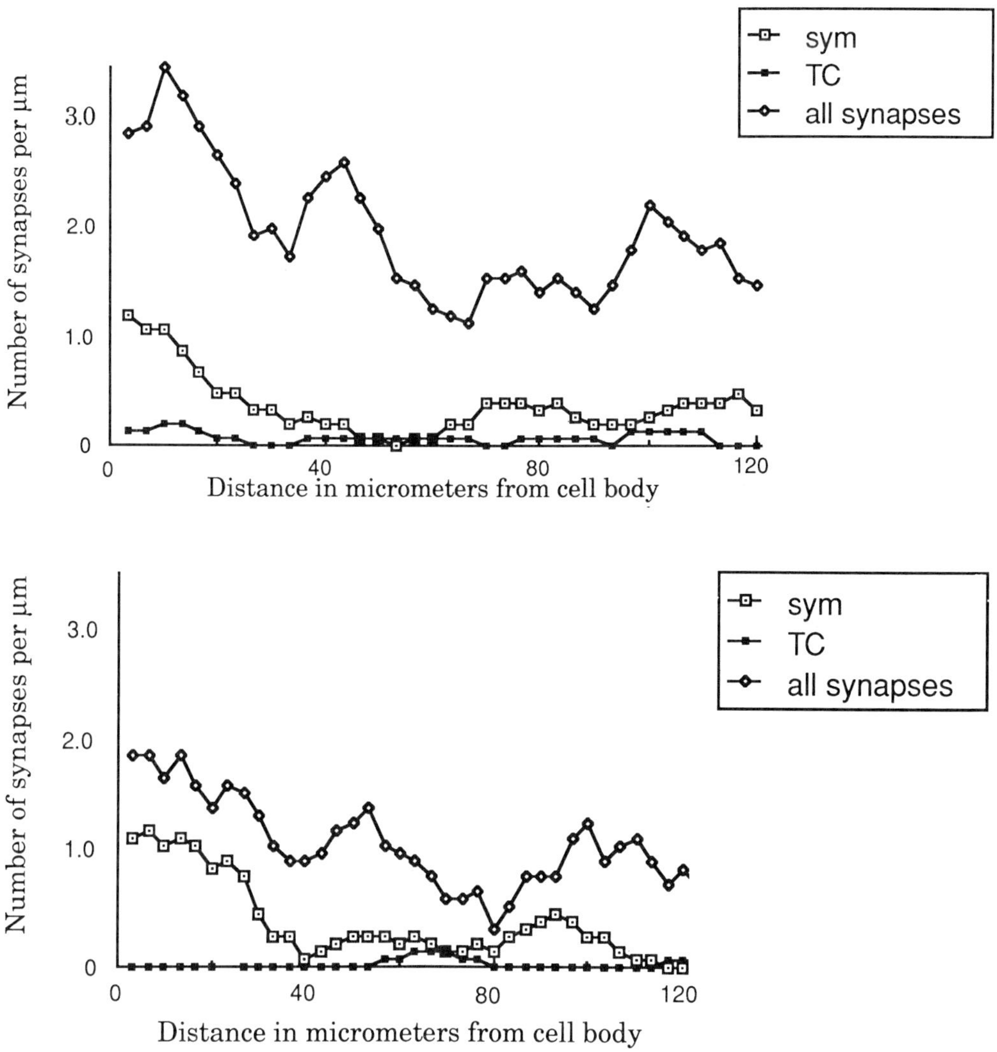

Figure 3.6. *Graphs plotting the spatial distribution of all synapses, symmetrical synapses, and thalamocortical synapses on two dendrites of a nonspiny multipolar neuron. Each point represents a boxcar integration for five 3-µm bins.*

other, but also on the concentration of the presynaptic elements in the neuropil. The dendrite in Figure 3.6b passes through neuropil close to the side of a layer IV barrel, whereas the dendrite shown in Figure 3.6a courses away from the side and runs almost entirely within the hollow of the barrel (White and Rock, 1981). Counts of synapses in barrels indicate that neuropil in barrel sides contains approximately 50% fewer synapses than neuropil in the barrel hollows; side neuropil contains a higher concentration of myelinated and unmyelinated axons and a correspondingly lower concentration of axon terminals (White, 1976). Thus, the dendrite of Figure 3.6b may receive fewer synapses than the dendrite of Figure 3.6a because the former passes through neuropil containing a lower concentration of presynaptic elements.

The same boxcar approach, in which distance from the cell body is considered the most important parameter, was used to plot the spatial distribution of synapses onto dendrites belonging to a nonspiny bitufted cell that occurred within the same barrel as the nonspiny multipolar neuron whose dendrites are the subject of Figure 3.6 (White and Rock, 1981). In general, the shapes of these curves parallel the curves depicted for dendrites of the multipolar cell: The curves for "all synapses" start off high and, with fluctuations, decrease with increasing distance from the cell body; the curves for symmetrical synapses fall off rapidly from a high point proximally and proceed roughly horizontally; the curves for thalamocortical synapses are consistently flat. Different results were obtained for a boxcar integration analysis of the distribution of synapses onto dendrites belonging to a layer IV spiny stellate cell that, like the nonspiny cells, is also from a barrel in the mouse primary somatosensory cortex (Figure 3.7) (White and Rock, 1980).

Figure 3.7 illustrates the boxcar integration for the longest dendrite of the spiny stellate cell. In this instance, the curve for "all synapses" rises from a low point, proceeds horizontally, and drops only toward the end of the dendrite; the curves for symmetrical and thalamocortical synapses are both flat. Plots for other dendrites of the same spiny stellate neuron were similar with the exception that the curves for symmetrical synapses were typically higher for the first few microns of the dendrite than in more distal regions. Although the data are few, it is tempting to speculate that the differently shaped "all synapses" curves (cf. Figure 3.6 with Figure 3.7) represent characteristic differences in the spatial

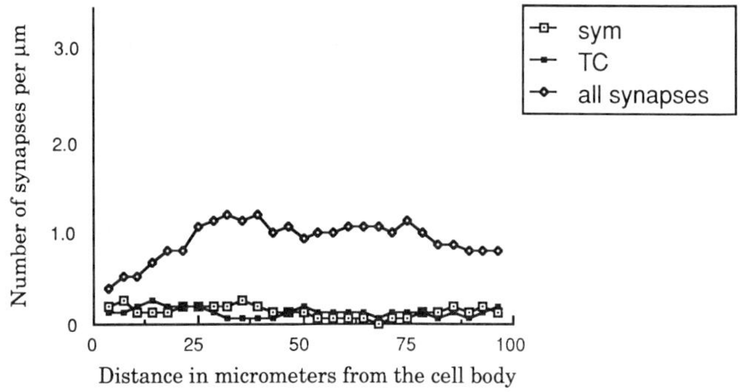

Figure 3.7. *Graph plotting the spatial distribution of all synapses, symmetrical synapses, and thalamocortical synapses on the longest dendrite of a spiny stellate cell. Each point represents a boxcar integration for five 3-µm bins.*

distribution of synapses onto nonspiny versus spiny, non-pyramidal neurons.

Additional evidence for differences in the spatial relationships of synapses onto spiny versus nonspiny dendrites is provided by analyses of thalamocortical synapses onto the layer IV spiny stellate neuron (White and Rock, 1980) and the layer IV nonspiny bitufted neuron mentioned above (White and Rock, 1981). For each of five dendrites belonging to the spiny stellate cell, the majority of intervals between spines receiving thalamocortical synapses was approximately 5 μm. In contrast, intervals between thalamocortical synapses were generally less than 2 μm for dendrites of the bitufted cell. Again, dendrites of the same neuron participate in similar synaptic arrangements, whereas dendrites of different types of neurons display different synaptic relationships.

Because more data are available for pyramidal than for nonpyramidal neurons, the question of specificity in the synaptic connectivities of dendrites will be taken up later on when the thalamocortical synaptic relationships of different types of pyramidal neurons is discussed. Still, it is worth mentioning that what quantitative data there are for nonpyramidal neurons is consistent with the idea that dendrites of different types of neurons receive characteristic synaptic patterns. For example, as mentioned earlier, dendrites of the nonspiny multipolar and bitufted neurons examined by White and Rock (1981) receive approximately 4% and 20%, respectively, of their asymmetrical synapses from thalamocortical afferents. All these dendrites passed through neuropil in which thalamocortical axon terminals accounted for approximately 20% of the synapses formed. Therefore, the fivefold difference in the frequency of thalamocortical synapses received by dendrites of the multipolar and bitufted neurons was not related to any corresponding difference in the concentration of thalamocortical synapses in the surrounding neuropil. Dendrites of the spiny stellate neuron examined by White and Rock (1979, 1980) received about 13% of their asymmetrical synapses from thalamocortical axon terminals, whereas several dendrites of a nonspiny multipolar neuron reconstructed from the same series of sections received 16% of their synapses from thalamocortical axon terminals (White, 1986). Here again, all dendrites passed through neuropil in which thalamocortical afferents formed approximately 20% of the synapses present.

The broad implication of these findings is that different neurons receive specific numbers and proportions of thalamocortical and other synapses. This specificity of synaptic circuitry contrasts sharply with the apparent nonselectivity with which thalamocortical axon terminals synapse with every type of dendrite in layer IV of sensory cortex.

The finding that, within one lamina, different dendrites of a single neuron have similar synaptic relationships provided the rationale for subsequent investigations of thalamocortical synaptic connectivity in which only selected dendrites or dendritic segments of nonpyramidal cells were examined. These studies include segments of dendrites belonging to 18 nonspiny multipolar neurons (Keller and White, 1987; White et al., 1984) and to 7 spiny stellate cells (Benshalom and White, 1986). In each report, the percentage of asymmetrical synapses formed by thalamocortical afferents was different for different dendrites: Extensive dendritic segments belonging to some nonspiny multipolar neurons received no thalamocortical synapses, and dendrites of others received from 1.5% to 21.3% of their asymmetrical synapses from thalamocortical afferents. The range for spiny stellate cells was much narrower, but, at 10.4% to 22.9%, still great enough to indicate that different spiny stellate cells receive markedly different numbers and proportions of thalamocortical synapses.

That different members of the same morphological cell type may form markedly different synaptic relationships suggests that morphological categories of nonpyramidal neurons, as presently defined, are likely to be composed of neurons that have very different functional roles. Consequently, caution should be exercised when making assumptions about the physiology of neuronal types on the basis of the shape of their cell body and on the distribution of their dendritic and axonal branches (see Part II for a more complete discussion of anatomical/functional correlations).

As mentioned in Chapter 2 (in the "Other Morphological Cell Types" section), further subdivision of the present morphological categories of nonpyramidal neurons may be possible by considering their content of neuroactive substances. For instance, in the hippocampus of the cat and in various areas of the cerebral cortex of the cat and monkey, some GABAergic nonpyramidal neurons contain cholecystokinin, whereas others contain somatostatin (Hendry et al., 1984; Somogyi et al., 1984). Because of the methods employed in these studies, dendritic trees were visualized only in part, and so the morphology of the neuronal types that contain these neuroactive substances is uncertain. However, on the basis of the available data, it seems highly probable that some overlap occurs in the morphological characteristics of neurons that have different chemical constituents. Thus, at some future time, it should be interesting to see whether the subdivision of nonpyramidal neuronal types on chemical grounds can be correlated with their thalamocortical synaptic relationships.

The thalamocortical synaptic connectivity of pyramidal neurons also has been the subject of investigations using quantitative approaches. As with nonpyramidal neurons, dendrites belonging to different pyramidal neurons may vary considerably in the numbers and proportions of thalamocortical synapses they receive, in spite of the fact that the neuropil surrounding each dendrite contains similar concentrations of thalamocortical synapses. For instance, Hersch and White (1981b) report that, in layer IV of the mouse SmI cortex, the proportion of thalamocortical synapses received by apical dendrites belonging to Golgi-impregnated, layer V and VI pyramidal neurons ranges from 1.3% to 14.6% of all the asymmetrical synapses formed onto these dendrites. Among the pyramidal cells of even a single cortical lamina are neurons that project to different target areas and that therefore have different functional roles (see Chapter 2).

The Golgi method provides no direct information on the functional roles of impregnated neurons, and so it cannot be determined if the range of values observed by Hersch and White (1981b) reflects a heterogeneity in the functional types of pyramidal neurons examined or whether the values represent some normal variation in the thalamocortical synaptic relationships of a single functional type of cortical pyramid. Some insight into this question has been obtained by using approaches to label pyramidal neurons on a somewhat more functionally oriented basis than is possible with the Golgi method, that is, by methods that enable the projection site of the pyramidal neuron to be identified.

Pyramidal cells can be classified according to the distant targets to which their axons project using relatively simple retrograde transport methods (see Chapter 2). Combining these methods with approaches designed to label synapses made by afferents to the cortex provides the possibility of correlating the synaptic input and output relationships of identified cortical neurons. In the present context, this issue can be phrased in the following way: *To what degree are the numbers and types of synaptic inputs different for pyramidal neurons that project to different regions of the brain, and to what degree are they similar for pyramidal neurons that have similar projections?* Insight into these questions has been provided by a series of quantitative studies of thalamocortical and corticocortical synapses in which the retrograde transport of horseradish peroxidase has been used to label pyramidal cells according to the site to which their axons project.

At present, quantitative studies of thalamocortical synapses with HRP-labeled projection neurons include examinations in layer IV of the mouse SmI cortex of basal dendrites belonging to superficial pyramidal cells that project from the SmI to the MsI cortex (White and Hersch, 1981) and of apical dendrites belonging to deep pyramidal cells that project from the SmI to either the ipsilateral striatum (Hersch and White, 1982), the ventrobasal thalamus, (Hersch and White, 1981c; White and Hersch, 1982), or to the MsI cortex (White et al., 1982). With the exception of dendrites belonging to deep corticocortical cells, all dendrites passed through fields of neuropil in which about 20% of the synapses were made by degenerating thalamocortical axon terminals. The apical dendrites of deep corticocortical pyramids passed through regions of layer IV in the mouse SmI cortex that contain few or no thalamocortical axon terminals (i.e., the walls of barrels; see Frost and Caviness, 1980; White, 1978; Woolsey and Dierker, 1978), and so these dendrites had essentially no opportunity to receive thalamocortical synapses. Dendrites belonging to each of the three other types of projection cell received thalamocortical synapses.

These findings, depicted graphically by the lower four bars in Figure 3.8, show that groups of pyramidal neurons that project to a single site (i.e., MsI cortex-CC^s, the striatum-CS, and the thalamus-CT) exhibit a range of thalamocortical synaptic connectivities. For comparison, the ranges of thalamocortical synaptic connectivities observed for Golgi-impregnated nonpyramidal neurons in the mouse SmI cortex are depicted in the upper two bars of Figure 3.8.

The only other quantitative study of identified synapses onto HRP-labeled projection neurons is an investigation of synapses between callosal afferents and callosal projection cells in the mouse MsI cortex (Porter and White, 1986). Findings show that apical dendrites belonging to callosal projection cells of layer V receive an average of less than half as many callosal synapses as compared with apical and basal dendrites belonging to callosal projection cells of layers II/III despite the fact that all the dendrites passed through neuropil containing comparable concentrations of callosal synapses. The percentage of callosal synapses onto dendrites of each type of projection neuron varied, but occurred within narrow ranges.

The ranges depicted in the lower half of Figure 3.8 for the thalamocortical synaptic connectivity of four projection cell types essentially do not overlap, although, in view of the small size of the samples, it is conceivable that some overlapping would occur were the sample size larger or were data included from additional types of projection cells. Notwithstanding these possibilities, the nonoverlapping ranges depicted

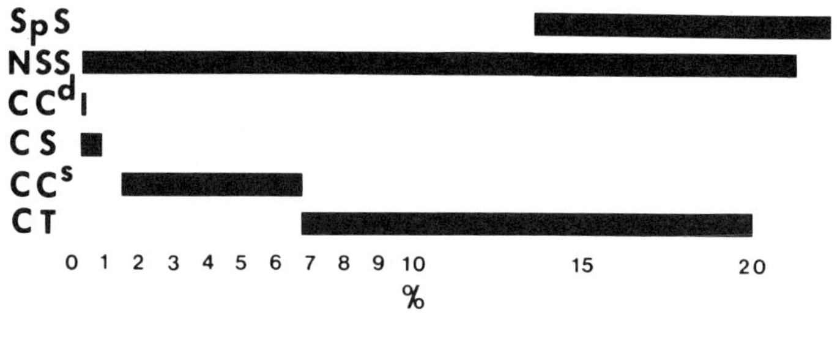

Figure 3.8. *Bar graph showing the percentage of synapses formed by thalamocortical axon terminals in layer IV of the mouse primary somatosensory cortex onto dendrites belonging to spiny stellate (SpS) neurons; nonspiny stellate (NSS) neurons; and to pyramidal neurons that project to the ipsilateral primary motor cortex (deep and superficial corticocortical cells, CC^d, CC^s), to the ipsilateral striatum (CS), and to the ipsilateral ventrobasal nucleus of the thalamus (CT). The CS, CC^d, and CT categories each contain 7 apical dendrites, whereas the CC^s category contains 12 basal dendrites from 6 cells. The SpS category includes 12 primary dendrites from 8 cells; the NSS category includes 14 primary dendrites and their branches from 10 cells. Apical dendrites of CC^d pyramids passed through regions of the cortex containing few or no thalamocortical axon terminals. All other dendrites occurred in neuropil where about 20% of the synapses were made by thalamocortical axon terminals (White, 1987).*

in Figure 3.8 are consistent the interpretation made by White and Hersch (1982) that specific types of projection neurons form characteristic proportions of synapses with thalamocortical afferents. Carried to its logical conclusion and taking into account data on the spatial relationships of synaptic inputs, this argument would suggest that specific types of pyramidal and nonpyramidal neurons receive characteristic numbers, types, and arrangements of synapses from both extrinsic and intrinsic sources.

The quantitative data presented in this section show that neurons within any morphological group exhibit a range of synaptic connectivities with extrinsic afferents. The basis for this heterogeneity is unclear, and several interpretations are possible. Some degree of heterogeneity can be expected in any biological system, and it may be that the differences observed are of slight consequence to brain function. However, as noted earlier, the differences observed in synaptic connectivity among members of a single morphological type may derive from including in each morphologically defined category neurons that have different functional roles within the cortex. An indication that this may well be the case with respect to projection cells is provided by findings of double-labeling studies showing that groups of pyramidal neurons that project to one site may contain some neurons that have additional projections to other sites (Adams et al., 1986; Goldman-Rakic and Schwartz, 1982; Kennedy and Bullier, 1985). Moreover, it may be that pyramidal neurons that share a common projection site differ in ways yet to be explored, such as with respect to the specific mix of neuroactive chemicals they contain.

Direct confirmation that neurons of the same morphological type may exhibit markedly different response properties has been adduced for both pyramidal and nonpyramidal neurons in the visual cortex of the rat and cat by the application of combined physiological recording and intracellular staining techniques (Kelly and Van Essen, 1974; Martin et al., 1983; Martin and Whitteridge, 1984; Parnevelas et al., 1983; Van Essen and Kelly, 1973; see also Chapter 4). These findings support the contention that a single morphological cell type is likely to contain a heterogeneous collection of neurons with regard to their synaptic connectivity and/or physiology. In view of this, findings showing in several instances that synaptic connectivities vary less within, than between, morphologically defined neuronal types are truly striking.

Summary

Extrinsic cortical afferents form most of their synapses with dendritic spines, but neuronal cell bodies and dendritic shafts are also contacted. Synapses with these elements are made in proportions that differ both according to the source of the afferent and according to the layer in which the synapses are made. Comparisons of the output relationships of different sets of afferents within the same cortical laminae indicate that extrinsic afferents are highly selective with regard to the kinds of targets with which they synapse.

Most quantitative information currently available on the synaptic connectivity of extrinsic afferents involves synapses made by thalamocortical afferents to mouse SmI cortex. In this area, each morphological neuronal type exhibits a range of values for the percentage of its synapses formed with thalamocortical afferents. This finding may indicate that a variety of functional types is included within each morphological category. Nevertheless, the ranges are sufficiently small to support the conclusion that specific neuronal types defined by their morphology or by their target projection sites form characteristic numbers, types, and spatial arrangements of thalamocortical and other synapses.

Summary of Chapter 3

Figure 3.9 is a summary diagram, which, in contrast with most other schemes of cortical connections, includes only those data that have been obtained by the visualization of synaptic connections with the electron microscope. This is not a mere technological distinction, for, as noted in the summary for Chapter 2, interpretations based on light microscopic studies have, on occasion, provided an erroneous impression of the circuitry actually present.

With regard to our understanding of cortical circuitry, perhaps the most damaging of these impressions is one relating to the synaptic connections of thalamocortical afferents with neurons in layer IV of the primary sensory cortices. The dendrites and cell bodies of spiny stellate cells occur primarily within layer IV of the primary sensory areas, the same lamina that contains the bulk of the specific thalamocortical projection to these regions (see the "Spiny Stellate Cells" section of Chapter 2). The coexistence of these elements within a single layer of the cortex has been misinterpreted to mean that spiny stellate cells are the nearly exclusive target of specific thalamic input to the cerebral cortex (cf. Eccles, 1984; and see the discussion of this point in Part III, Chapter 7). This impression, however, is nothing more than an illusion derived from the erroneous interpretation of data obtained with the light microscope. Moreover, the findings of quantitative studies indicate that both corticothalamic projection cells, despite the paucity of their dendritic branches in layer IV, and certain nonspiny nonpyramidal neurons, receive many thalamocortical synapses. Therefore, it cannot be excluded that a majority of thalamocortical synapses are formed with types of neurons other than spiny stellate cells.

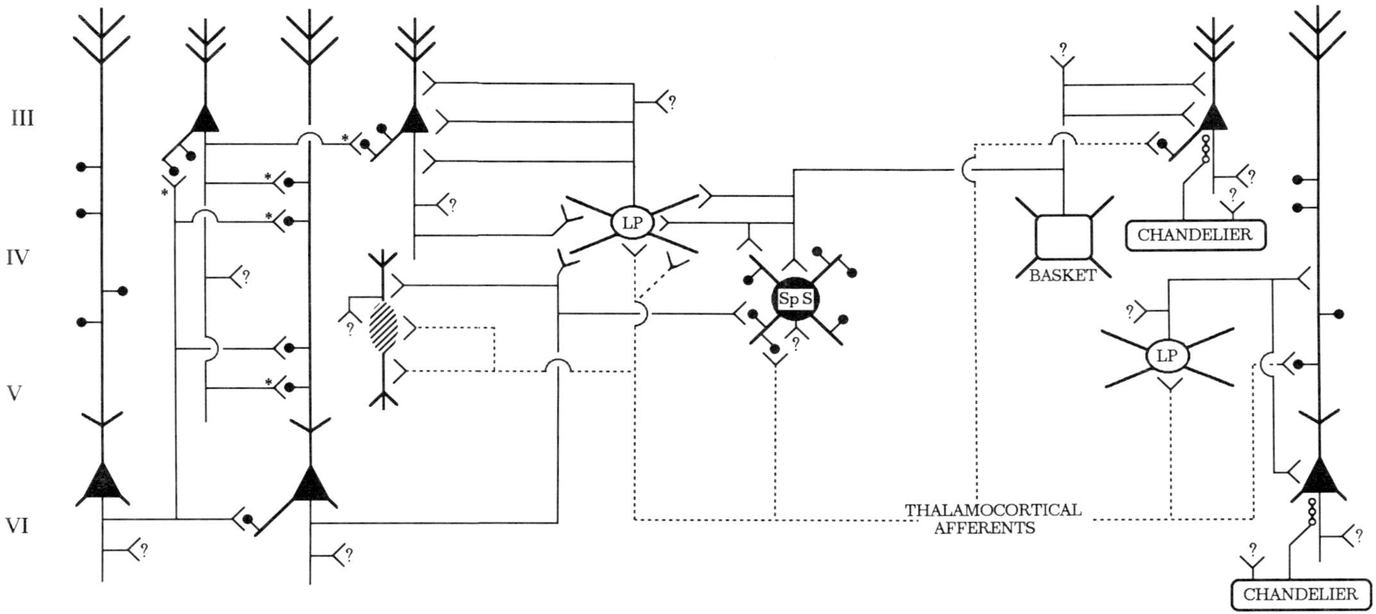

Figure 3.9. *Summary diagram illustrating the synaptic connections in cortex that have been conclusively demonstrated by labeling both their pre- and postsynaptic elements in single preparations. Also shown are several synapses (*) whose existence is highly likely (see the "Intrinsic Connections of Pyramidal Neurons" section). Neurons whose axons form asymmetrical synapses, that is pyramidal neurons and spiny stellate cells (SpS), are shown in solid black; neurons whose axons form symmetrical synapses, that is local plexus (LP), basket, and chandelier cells, are unfilled. The two varieties of nonspiny bipolar cells, one having an axon that forms asymmetrical synapses and the other an axon that forms symmetrical ones, are shown by a single crosshatched cell. Thalamocortical afferents are shown by dashed lines.*

General Aspects of Brain Circuitry

In line with the premise that all regions of the cerebral cortex are organized according to certain basic principles, the following generalizations of specific findings presented in Part I are offered as basic rules governing the synaptic organization of the cerebral cortex:

Rule 1. Every neuron within the target region of a projection receives input from the projection.

Corollary to Rule 1. Axon terminals from any extrinsic or intrinsic source synapse onto every morphological or physiological neuronal type within their terminal projection field. In practice, this means that a pathway will form synapses with every element in its target region capable of forming the type of synapse normally made by the pathway (i.e., asymmetrical or symmetrical).

Rule 2. Different dendrites of a single neuron form similar synaptic patterns; that is, the numbers, types, proportions, and spatial distribution of synapses is similar, provided the dendrites are exposed to similar synaptic inputs.

Corollary to Rule 2. Axonal pathways form similar synaptic patterns onto all the dendrites of a single neuron, provided the dendrites occur within the target region of the axonal pathway.

Rule 3. Neuronal types receive characteristic patterns of synaptic connections; the actual numbers, proportions, and spatial distribution of the synapses formed by each neuronal type occur within a range of values.

Corollary to Rule 3. Different extrinsic and intrinsic synaptic pathways form specific proportions of their synapses with different postsynaptic elements (spines vs. dendritic shafts, one cell type vs. another).

References for Part I

Abeles, M., and Goldstein, M.H., Jr. (1970). Functional architecture in cat primary auditory cortex: Columnar organization and organization according to depth. *J. Neurophysiol.* *33,* 172–187.

Adams, C.E., Cepeda, C., Boylan, M.K., Fisher, R.S., Hull, C.D., Buchwald, N.A., Wainer, B.H., and Levine, M.S. (1986). Basal forebrain neurons have axon collaterals that project to widely divergent cortical areas in the cat. *Brain Res.* *397,* 365–371.

Ades, H.W. (1959). Central auditory mechanisms. In: *Handbook of Physiology,* vol. 1 (J. Field, ed.). American Physiology Society, Washington, D.C., pp. 585–613.

Asanuma, H. (1975). Recent developments in the study of the columnar arrangement of neurons within the motor cortex. *Physiol. Rev.* *55,* 143–156.

Asanuma, H., and Rosén, I. (1972). Topographical organization of cortical efferent zones projection to distal forelimb muscles in the monkey. *Exp. Brain Res.* *14,* 243–256.

Aston-Jones, G., Foote, S.L., and Bloom, F.E. (1984). Anatomy and physiology of locus coeruleus neurons: Functional implications. In: *Norepinephrine* (M.G. Ziegler and C.R. Lake, eds.). Williams and Wilkins, Baltimore, pp. 92–116.

Bailey, P., and von Bonin, G. (1951). *The Isocortex of Man.* Univ. of Illinois Press, Urbana. Cited in Braak (1984).

Ballard, D.H., Hinton, G.E., and Sejnowski, T.J. (1983). Parallel visual computation. *Nature* *306,* 21–26.

Barbaresi, P., Fabri, M., Conti, S., and Manzoni, T. (1987). D'(3H) aspartate retrograde labeling of callosal and asociation neurons of somatosensory areas I and II of cats. *J. Comp. Neurol.* *263,* 159–178.

Bartfai, T. (1985). Presynaptic aspects of the coexistence of classical neurotransmitters and peptides. *T.I.P.S.,* August, 331–333.

Bartfai, T., Iverfeldt, K., Brodin, E., and Ögren, S.-O. (1986). Functional consequences of coexistence of classical and peptide neurotransmitters. In: *Progress in Brain Research,* vol. 68 (T. Hökfelt, K. Fuxe, and B. Pernow, eds.). Elsevier, Amsterdam, pp. 321–330.

Baughman, R.W., and Gilbert, C.D. (1981). Aspartate and glutamate as possible neurotransmitters of cells in layer 6 of the visual cortex. *Nature 287,* 848–849.

Bear, M.F., Carnes, K.M., and Ebner, F.F. (1985) An investigation of cholinergic circuitry in cat striate cortex using acetylcholinesterase histochemistry. *J. Comp. Neurol. 234,* 411–430.

Bear, M.F., and Singer, W. (1986). Modulation of visual cortical plasticity by acetylcholine and noradrenaline. *Nature 320,* 172–176.

Benshalom, G., and White, E.L. (1986). Quantification of thalamocortical synapses with spiny stellate neurons in layer IV of mouse somatosensory cortex. *J. Comp. Neurol. 253,* 303–314.

Blackstad, T.W. (1965). Mapping of experimental axon degeneration by electron microscopy of Golgi preparations. *Z. Zellforsch. Mikros. Anat. 67,* 819–834.

Bloom, F.E. (1981). Chemical signaling and cortical circuitry: Integrative aspects. In: *The Organization of the Cerebral Cortex* (G. Adelman, S.G. Dennis, F.O. Schmitt, and F.G. Worden, eds.), MIT Press, Cambridge, Mass., pp. 359–370.

Bodian, D. (1952). Introductory survey of neurons. *Cold Spring Harbor Symposia on Quantitative Biology,* vol. 17, p. 1–13.

Bolton J.S. (1910). A contribution to the localization of cerebral function, based on the clinicopathological study of mental disease. *Brain 33,* 26–147.

Bormann, J. (1988). Electrophysiology of $GABA_A$ and $GABA_B$ receptor subtypes. *T.I.N.S. 11,* 112–116.

Braak, H. (1976). On the striate area of the human isocortex: A Golgi and pigmentarchitectonic study. *J. Comp. Neurol. 166,* 341–364.

Braak, H. (1984). Architectonics as seen by lipofuchsin stains. In: *The Cerebral Cortex,* Vol. 1, The Cellular Components of the Cerebral Cortex (E.G. Jones and A. Peters, eds.), Plenum Press, New York, pp. 59–104.

Braitenberg, V. (1977a). The cellular components of the cerebral cortex. In: *Lecture Notes in Biomathematics,* vol. 21 (R. Heim and G. Palm, eds.). Springer, Berlin, pp. 171–188.

Braitenberg, V. (1977b). *On the Texture of Brains.* Springer-Verlag, New York.

Braitenberg, V. (1978). Cortical architectonics: General and areal. In: *Architectonics of the Cerebral Cortex* (M.A.B. Brazier and H. Petsche eds.). Raven Press, New York, pp. 443–465.

Braitenberg, V. (1985). Charting the visual cortex. In: *The Cerebral Cortex.* Vol. 3, Visual Cortex (A. Peters and E.G. Jones, eds.). Plenum Press, New York, pp. 379–414.

Brodmann, K. (1906). Beitrage zur histologischen Lokalisation der Grosshirnrinde: Über den allgemeinen Bauplan des Cortex pallii bei den Mammaliern und zwei homologe Rindenfelder im Besonderen. Zugleich ein Beitrag zur Furschenlehre. *J. Psychol. Neurol. 6,* 275. Cited in Kemper and Galaburda (1984).

Brodmann, K. (1909). Vergleichende Lokalisationslehre der Grosshirnrinde in Ihren Prinzipien dargestellt auf Grund des Zellenbaues. Barth, Leipzig. Cited in Kemper and Galaburda (1984).

Brugge, J.F., and Reale, R.A. (1985). Auditory cortex. In: *The Cerebral Cortex.* Vol 4, Association and Auditory Cortices (A. Peters and E.G. Jones, eds.). Plenum Press, New York. pp. 229–271.

Cajal, R.S. (1909–1911). *Histologie du système nerveux de l'homme et des vertebrès,* vol. II. Paris: Maloine.

Campbell, A.W. (1905). *Histological Studies on the Localization of Cerebral Function.* Cambridge University Press, London.

Campos, G.B., and Welker, W.I. (1976). Comparisons between brains of a large and small hystricomorph rodent: Capybara, Hydochoerus and guinea pig, Cavia; Neocortical projection regions and measurements of brain subdivisions. *Brain Behav. and Evol.* 13, 243–266.

Chapin, J.K. (1986). Laminar differences in sizes, shapes, and response profiles of cutaneous receptive fields in the rat SI cortex. *Exp. Brain Res. 62,* 549–559.

Chmielowska, J., and Kossut, M. (1984). A single vibrissal column in the first somatosensory cortex of the mouse demonstrated with 2-deoxyglucose. *Acta. Neurobiol. Exp. 44,* 83–88.

Cipolloni, P.B., Hersch, S.M., and White, E.L. (1985). The use of lectin transport as an anterograde marker for electron microscopy. *Neurosci. Lett. 58,* 43–47.

Cipolloni, P.B., and Peters, A. (1983). The termination of callosal fibres in the auditory cortex of the rat. A combined Golgi-electron microscope and degeneration study. *J. Neurocytol. 12,* 713–726.

Code, R.A., and Winer, J.A. (1985). Commisural neurons in layer III of cat primary auditory cortex (AI): Pyramidal and non-pyramidal cell input. *J. Comp. Neurol. 242,* 485–510.

Colonnier, M. (1964). Experimental degeneration in the cerebral cortex. *J. Anat. 98,* 47–53.

Colonnier, M. (1968). Synaptic patterns on different cell types in the different laminae of the cat visual cortex. An electron microscope study. *Brain Res. 9,* 268–287.

Colonnier, M. (1981). The electron-microscopic analysis of the neuronal organization of the cerebral cortex. In: *The Organization of the Cerebral Cortex* (G. Adelman, S.G. Dennis, F.O. Schmitt, and F.G. Worden, eds.). MIT Press, Cambridge, Mass., 125–153.

Conti, F., Fabri, M., and Manzoni, T. (1987). An immunocytochemical study on glutaminergic cortico-cortical neurons in the somatic sensory areas of monkeys. *Society for Neuroscience Abstracts of Annual Meeting,* vol. 13. no. 73.16, p. 250.

Coss, R.G., and Perkel, D.H. (1985). The function of dendritic spines: A review of theoretical issues. *Behav. and Neurol. Biol. 44,* 151–185.

Crawford, M.L.J. (1985). Stimulus-specific columns in monkey visual cortex as revealed by the [^{14}C]-2-deoxyglucose method. In: *The Cerebral Cortex.* Vol. 3, Visual Cortex (A. Peters and E.G. Jones, eds.). Plenum Press, New York, pp. 331–349.

Creutzfeldt, O.D. (1977). Generality of the functional structure of the neocortex. *Naturwissenschaften 64,* 507–517.

Creutzfeldt, O.D. (1978). The neocortical link: Thoughts on the generality of structure and function of the neocortex. In: *Architectonics of the Cerebral Cortex* (M.A.B. Brazier and H. Petsche, eds.). Raven Press, New York, pp. 357–391.

Cuenod, M., Bagnoli, P., Beaudet, A., Rustiono, A., Wiklund, L., and Streit, P. (1982). Transmitter-specific retrograde labeling of neurons. In: *Cytochemical Methods in Neuroanatony,* vol. 1, V (V. Chan-Palay and S.L. Palay, eds.). Alan R. Liss, New York, pp. 17–44.

Cunningham, E.T., Jr., and LeVay, S. (1986). Laminar and synaptic organization of the projection from the thalamic nucleus centralis to primary visual cortex in the cat. *J. Comp. Neurol. 254,* 65–77.

Curcio, C.A., and Harting, J.K. (1978). Organization of pulvinar afferents to area 18 in the squirrel monkey: Evidence for stripes. *Brain Res. 143,* 155–161.

Davis, T.L., and Sterling, P. (1979). Microcircuitry of cat visual cortex: Classification of neurons in layer IV of area 17, and identification of the patterns of lateral geniculate input. *J. Comp. Neurol. 188,* 599–628.

DeCarlos, J.A., Lopez-Mascaraque, L., Ramon y Cajal-Agueras, S., and Valverde, F. (1987). Chandelier cells in the auditory cortex of monkey and man: A Golgi study. *Exp. Brain Res. 66,* 295–302.

DeCarlos, J.A., Lopez-Mascaraque, L., and Valverde, F. (1985). Development, morphology and topography of chandelier cells in the auditory cortex of the cat. *Dev. Brain Res. 22,* 293–300.

DeFelipe, J., Conley, J., and Jones, E.G. (1986a). Long-range focal collateralization of axons arising from corticocortical cells in monkey sensory-motor cortex. *J. Neurosci. 6,* 3749–3766.

DeFelipe, J., and Fairén, A. (1982). A type of basket cells in superficial layers of the cat visual cortex. A Golgi-electron microscope study. *Brain Res. 244,* 9–16.

DeFelipe, J., Hendry, S.H.C., and Jones, E.G. (1986b). A correlative electron microscopic study of basket cells and large GABAergic neurons in the monkey sensory-motor cortex. *Neuroscience 17,* 991–1009.

DeFelipe, J., Hendry, S.H.C., Jones, E.G., and Schmechel, D. (1985). Variability in the terminations of GABAergic chandelier cell axons on initial segments of pyramidal cell axons in the monkey sensory-motor cortex. *J. Comp. Neurol. 231,* 364–384.

Demeulemeester, H., Vandesande, F., Orban, G.A., Brandon, C., and Vanderhaeghen, J.J. (1988). Heterogeneity of GABAergic cells in cat visual cortex. *J. Neurosci. 8,* 988–1000.

Dreher, B., Leventhal, A.G., and Hale, P.T. (1980). Geniculate input to cat visual cortex: A comparison of area 19 with areas 17 and 18. *J. Neurophysiol. 44,* 804–826.

Durham, D., and Woolsey, T.A. (1977). Barrels and columnar organization: Evidence from 2-deoxy-glucose (2-DG) experiments. *Brain Res. 137,* 169–174.

Durham, D., and Woolsey, T.A. (1978). Acute whisker removal reduces neuronal activity in barrel of mouse SmI cortex. *J. Comp. Neurol. 178*, 629–644.

Durham, D., and Woolsey, T.A. (1985). Functional organization in cortical barrels of normal and vibrissae-damaged mice: A [3H]2-deoxyglucose study. *J. Comp. Neurol. 235*, 97–110.

Dykes, R.W. (1978). The anatomy and physiology of the somatic sensory cortical regions. *Prog. Neurobiol., 10*, 33–88.

Dykes, R.W. (1983). Parallel processing of somatosensory information: A theory. *Brain Res. Rev.* 6, 47–115.

Dykes, R.W., and Ruest, A. (1986). What makes a map in somatosensory cortex? In: *The Cerebral Cortex.* Vol. 5, Sensory-motor Areas and Aspects of Cortical Connectivity. (E.G. Jones and A. Peters, eds.). Plenum Press, New York, pp. 1–29.

Eccles, J.C. (1984). The cerebral neocortex: A theory of its operation. In: *The Cerebral Cortex.* Vol. 2, Functional Properties of Cortical Cells (E.G. Jones and A. Peters, eds.). Plenum Press, New York, pp. 1–36.

Einstein, G., Davis, T.L., and Sterling, P. (1987). Ultrastructure of synapses from the A-laminae of the lateral geniculate nucleus in layer IV of the cat striate cortex *J. Comp. Neurol. 260*, 63–75.

Emson, P.C., and Hunt, S.C. (1984). Peptide-containing neurons of the cerebral cortex. In: *The Cerebral Cortex.* Vol. 2, Functional Properties of Cortical Cells (E.G. Jones and A. Peters, eds.). Plenum Press, New York, pp. 145–169.

Fairén, A., DeFelipe, J., and Martínez-Ruiz, R. (1981). The Golgi-EM procedure: A tool to study neocortical interneurons. In: *Eleventh International Congress of Anatomy: Glial and Neuronal Cell Biology.* Alan R. Liss, New York, pp. 291–301.

Fairén, A., DeFelipe, J., and Regidor, J. (1984). Nonpyramidal neurons: General account. In: *The Cerebral Cortex.* Vol 1, Cellular Components of the Cerebral Cortex (A. Peters and E.G. Jones, eds.). Plenum Press, New York, pp. 201–253.

Fairén, A., Peters, A., and Saldanha, J. (1977). A new procedure for examining Golgi impregnated neurons by light and electron microscopy. *J. Neurocytol.* 6, 311–337.

Fairén, A., and Valverde, F. (1980). A specialized type of neuron in the visual cortex of cat: A Golgi and electron microscope study of chandelier cells. *J. Comp. Neurol. 194*, 761–779.

Fallon, J.H., and Loughlin, S.E. (1987). Monoamine innervation of cerebral cortex and a theory of the role of monoamines in cerebral cortex and basal ganglia. In: *The Cerebral Cortex.* Vol. 6, Further Aspects of Cortical Function, Including Hippocampus (E.G. Jones and A. Peters, eds.). Plenum Press, New York, pp. 41–127.

Feldman, M.L. (1984). Morphology of the neocortical pyramidal neuron. In: *The Cerebral Cortex.* Vol 1, Cellular Components of the Cerebral Cortex (A. Peters and E.G. Jones, eds.). Plenum Press, New York, pp. 123–200.

Feldman, M.L., and Peters, A. (1978). The forms of non-pyramidal neurons in the visual cortex of the rat. *J. Comp. Neurol. 179*, 761–794.

Ferster, D., and Lindström, S. (1985). Synaptic excitation of neurones in area 17 of the cat by intracortical axon collaterals of corticogeniculate cells. *J. Physiol.* (London) *367,* 233–252.

Fisken, R.A., Garey, L.J., and Powell, T.P.S. (1975). The intrinsic, association and commissural connections of area 17 of the visual cortex. *Phil. Trans. R. Soc.* (London) *B 272,* 487–536.

Fleischauer, K. (1978). Cortical architectonics: The last 50 years and some problems of today. In: *Architectonics of the Cerebral Cortex* (M.A.B. Brazier and H. Petsche, eds.). Raven Press, New York, pp. 99–117.

Foote, S.L. and Morrison, J.L. (1987). Extrathalamic Modulation of Cortical Function. *Ann. Rev. Neurosci. 10,* 67–95.

Freund, T.F., Magloczky, Z., Soltesz, I., and Somogyi, P. (1986). Synaptic connections, axonal and dendritic patterns of neurons immunoreactive for cholecystokinin in the visual cortex of the cat. *Neuroscience 19,* 1133–1159.

Freund, T.F., Martin, K.A.C., Smith, A.D., and Somogyi, P. (1983). Glutamate decarboxylase-immunoreactive terminals of Golgi-impregnated axoaxonic cells and of presumed basket cells in synaptic contact with pyramidal neurons of the cat's visual cortex. *J. Comp. Neurol. 221,* 263–278.

Freund, T.F., Martin, K.A.C., Somogyi, P., and Whitteridge, D. (1985). Innervation of cat visual areas 17 and 18 by physiologically identified X- and Y- type thalamic afferents. II. Identification of postsynaptic targets by GABA immunocytochemistry and Golgi impregnation. *J. Comp. Neurol. 242,* 275–291.

Friedman, D.P. (1983). Laminar patterns of termination of cortico-cortical afferents in the somatosensory system. *Brain Res. 273,* 147–151.

Frost, D.O., and Caviness, V.S., Jr. (1980). Radial organization of thalamic projections to the neocortex in the mouse. *J. Comp. Neurol. 194,* 369–393.

Gabbott, P.L.A., Martin, K.A.C., and Whitteridge, D. (1987). Connections between pyramidal neurons in layer 5 of cat visual cortex (area 17). *J. Comp. Neurol. 259,* 364–381.

Galaburda, A.M., and Pandya, D.N. (1983). The intrinsic architectonic and connectional organization of the superior temporal region of the rhesus monkey. *J. Comp. Neurol. 221,* 169–184.

Gamrani, H., Harandi, M., Dubois, M.P., and Calas, A. (1984). Direct electron microscopic evidence for the coexistence of GABA uptake and endogenous serotonin in the same rat central neurons by coupled radioautographic and immunocytochemical procedures. *Neurosci. Lett. 48,* 25–30.

Garbarg, M., Barbin, G., Llorens, C., Palacios, J.M., Pollard, H., and Schwatz, J. C. (1980). Recent developments in brain histamine research: Pathways and receptors. In: *Neurotransmitters, Receptors and Drug Action* (W.B. Essman, ed.). Spanish Medical and Scientific Books, Jamaica, pp. 179–202.

Garey, L.J. (1971). A light and electron microscopic study of the visual cortex of the cat and monkey. *Proc. Roy. Soc. Lond. Ser. B. 179,* 21–40.

Garey, L.J., and Powell, T.P.S. (1971). An experimental study of the termination of the lateral geniculo-cortical pathway in the cat and monkey. *Proc. Roy. Soc.* ser. B, *179,* 21–40.

Gennari, F. (1782). De Peculiari Structura Cerebri Nonnulisque Ejus Morbis. Parma. In: Brazier, M.A.B. (1978). Architects of the Cerebral Cortex: Research in the 19th Century. In: *Architectonics of the Cerebral Cortex* (M.A.B. Brazier and H. Petsche, eds.). Raven Press, New York, pp. 9–29.

Gilbert, C.D. (1977). Laminar differences in receptive field properties in cat primary visual cortex. *J. Physiol. 268,* 391–421.

Gilbert, C.D., and Kelly, J.P. (1975). The projections of cells in different layers of the cat's visual cortex. *J. Comp. Neurol. 163,* 81–106.

Gilbert, C.D., and Wiesel, T.N. (1979). Morphology and intracortical projections of functionally identified neurons in the cat visual cortex. *Nature 280,* 120–125.

Gilbert, C.D., and Wiesel, T.N. (1981). Laminar specialization and intracortical connections in cat primary visual cortex. In: *The Organization of the Cerebral Cortex* (G. Adelman, S.G. Dennis, F.O. Schmidt, and F.G. Worden, eds.). MIT Press, Cambridge, Mass., pp. 163–191.

Globus, A., and Scheibel, A.B. (1967). Synaptic loci on visual cortical neurons of the rabbit: The specific afferent radiation. *Exp. Neurol. 18,* 116–131.

Goldman, P.S., and Nauta, W.J.H. (1977). Columnar distribution of corticocortical fibers in the frontal association, limbic and motor cortex of the developing rhesus monkey. *Brain Res. 122,* 393–413.

Goldman-Rakic, P.S., and Schwartz, M.L. (1982). Interdigitation of contralateral and ipsilateral columnar projections to frontal association cortex in primates. *Science 216,* 755–757.

Gonzalez, M.F., and Sharp, F.R. (1985). Vibrissae tactile stimulation: (14C)2-deoxyglucose uptake in rat brainstem, thalamus and cortex. *J. Comp. Neurol. 231,* 457–472.

Gordon, B., Allen, E.E., and Trombley, P.Q. (1988). The role of neorepinephrine in plasticity of visual cortex. *Prog. Neurobiol. 30,* 171–191.

Gray, E.G. (1959). Axo-somatic and axodendritic synapses of the cerebral cortex: An electron microscopic study. *J. Anat.* (London) *93,* 420–433.

Gruner, J.E., Hirsch, J.C., and Sotelo, C. (1974). Ultrastructural features of the isolated suprasylvan gyrus. *J. Comp. Neurol. 154,* 1–27.

Gudden, B. (1870). Experimentaluntersuchungen über das Peripherische und centrale Nervensystem. *Arch. Psychiat. 2,* 693–723. Cited in A.E. Walker, *The Primate Thalamus,* Univ. of Chicago Press, Chicago, 1938.

Haberly, L.B. (1983). Structure of the piriform cortex of the opossum. I. Description of neuron types with Golgi methods. *J. Comp. Neurol. 213,* 163–187.

Haberly, L.B., and Feig, S. (1983). Structure of the piriform cortex of the opossum. II. Fine structure of cell bodies and neuropil. *J. Comp. Neurol. 216,* 69–88.

Haberly, L.B., and S. Presto (1986). Ultrastructural analysis of synaptic relationships of intracellularly stained pyramidal cell axons in piriform cortex. *J. Comp. Neurol. 248,* 464–474.

Hand, P.J., Juliano, S., Whitsel, B.G., and Goochee, C. (1979). A (14-C)-deoxyglucose study of the modular organization in somatosensory cortices of the cynomologous monkey. *Anat. Rec., 193,* 560–561.

Harris, R.M., and Woolsey, T.W. (1983). Computer-assisted analyses of barrel neuron axons and their putative synaptic contacts. *J. Comp. Neurol. 220,* 63–79.

Haug, H. (1987). Brain sizes, surfaces, and neuronal sizes of the cortex cerebri: A stereological investigation of man and his variability and a comparison with some mammals (primates, whales, marsupials, insectivores, and one elephant). *Amer. J. Anat. 180.* 126–142.

Hedlich, A., and Winkelmann, E. (1982). Neuronentypen des visuellen cortex der adulten und juvenilen ratte. *J. Hirnforsch. 23,* 353–373.

Hendrickson, A. (1972). Electron microscopic distribution of axoplasmic transport. *J. Comp. Neurol. 144,* 381–398.

Hendrickson, A. (1975). Technical modifications to facilitate tracing synapses by electron microscopic autoradiography. *Brain Res. 85,* 241–247.

Hendrickson, A., Hunt, S.P., and Wu, J.-Y. (1981). Immunocytochemical localization of glutamic acid decarboxylase in monkey striate cortex. *Nature* (London) *292,* 605–607.

Hendry, S.H.C., and Jones, E.G. (1980). Electron microscopic demonstration of thalamic axon terminations on identified commissural neurons in monkey somatic sensory cortex. *Brain Res. 196,* 253–257.

Hendry, S.H.C., and Jones, E.G. (1981). Size and distribution of neurons incorporating tritiated GABA in monkey sensory-motor cortex. *J. Neurosci. 1,* 390–408.

Hendry, S.H.C., and Jones, E.G. (1983). Thalamic inputs to identified commissural neurons in the monkey somatic sensory cortex. *J. Neurocytol. 12,* 299–316.

Hendry, S.H.C., Jones, E.G., DeFelipe, J., Schmechel, D., Brandon, C., and Emson, P.C. (1984). Neuropeptide-containing neurons of the cerebral cortex are also GABAergic. *Proc. Natl. Acad. Sci. USA 81,* 6526–6530.

Hersch, S.M., and White, E.L. (1981a). Thalamocortical synapses involving identified neurons in mouse primary somatosensory cortex: A terminal degeneration and Golgi/EM study. *J. Comp. Neurol. 195,* 252–263.

Hersch, S.M., and White, E.L. (1981b). Quantification of synapses formed with apical dendrites of Golgi impregnated pyramidal cells: Variability in thalamocortical inputs and consistency in the ratios of asymmetrical to symmetrical synapses. *Neuroscience 6,* 1043–1051.

Hersch, S.M., and White, E.L. (1981c). Thalamocortical synapses with corticothalamic projection neurons in mouse SmI cortex: Electron microscopic demonstration of a monosynaptic feedback loop. *Neurosci. Lett. 24,* 207–210.

Hersch, S.M., and White, E.L. (1982). A quantitative study of the thalamocortical and other synapses in layer IV of pyramidal cells projecting from mouse SmI cortex to the caudate-putamen nucleus. *J. Comp. Neurol.*

Hornung, J.P., and Garey, L.J. (1980). A direct pathway from the thalamus to visual callosal neurons in cat. *Exp. Brain Res. 38,* 121–123.

Hornung, J.P., and Garey, L.J. (1981). The thalamic projection to cat visual cortex: Ultrastructure of neurons identified by Golgi impregnation or retrograde horseradish peroxidase transport. *Neuroscience 6,* 1053–1068.

Houser, C.R., Hendry, S.H.C., Jones, E.G., and Vaughn, J.E. (1983). Morphological diversity of immunocytochemically identified GABA neurons in the monkey sensory-motor cortex. *J. Neurocytol.* *12,* 617–638.

Houser, C.R., Vaughn, J.E., Hendry, S.H.C., Jones, E.G., and Peters, A. (1984). GABA neurons in the cerebral cortex. In: *The Cerebral Cortex.* Vol. 2, Functional Properties of Cortical Cells (E.G. Jones and A. Peters, eds.). Plenum Press, New York, pp. 63–89.

Hubel, D.H., and Wiesel, T.N. (1962). Receptive fields, binocular interaction and functional architecture in the cat's visual cortex. *J. Physiol.* (London) *160,* 106–154.

Hubel, D.H., and Wiesel, T.N. (1968). Receptive fields and functional architecture of monkey striate cortex. *J. Physiol.* *195,* 215–243.

Hubel, D.H., and Wiesel, T.N. (1972). Laminar and columnar distribution of geniculocortical fibers in the macaque monkey. *J. Comp. Neurol.* *146,* 421–450.

Hubel, D.H., and Wiesel, T.N. (1977). Functional architecture of macaque monkey visual cortex. *Proc. Roy. Soc.* (London) ser. B. 198, 1–59.

Hubel, D.H., Wiesel, T.N., and Stryker, P.M. (1978). Anatomical demonstration of orientation columns in macaque monkey. *J. Comp. Neurol.* *177,* 361–380.

Ichikawa, M., Arissian, K., and Asanuma, H. (1985). Distribution of corticocortical and thalamocortical synapses on identified motor cortical neurons in the cat: Golgi, electron microscopic and degeneration study. *Brain Res.* *345,* 87–101.

Innocenti, G.M. (1986). General organization of callosal connections in the cerebral cortex. In: *The Cerebral Cortex.* Vol. 5, Sensory-motor Areas and Aspects of Cortical Connectivity (E.G. Jones and A. Peters, eds.). Plenum Press, New York, pp. 291–353.

Innocenti, G.M., and Fiore, L. (1976). Morphological correlates of visual field transformation in the corpus callosum. *Neurosci. Lett.* *2,* 245–252.

Jacobson, S., and Trojanowski, J.Q. (1975). Corticothalamic neurons and thalamocortical terminal fields: An investigation in rat using horseradish peroxidase and autoradiography. *Brain Res.* *85,* 385–401.

Jacobson, S., and Trojanowski, J.Q. (1977a). Prefrontal granular cortex of the rhesus monkey. I. Intrahemispheric cortical afferents. *Brain Res.* *132,* 209–233.

Jacobson, S., and Trojanowski, J.Q. (1977b). Prefrontal granular cortex of the rhesus monkey. II. Interhemispheric cortical afferents. *Brain Res.* *132,* 235–246.

Jones, E.G. (1975). Varieties and distribution of non-pyramidal cells in the somatic sensory cortex of the squirrel monkey. *J. Comp. Neurol.* *160,* 205–268.

Jones, E.G., Coulter, J.D., Burton, H., and Porter, R. (1977). Cells of origin and terminal distribution of corticostriatal fibers arising in the sensory-motor cortex of monkeys. *J. Comp. Neurol.* *173,* 53–80.

Jones, E.G., Coulter, J.D., and Wise, S.P. (1979). Commissural columns in the sensory-motor cortex of monkeys. *J. Comp. Neurol.* *188,* 113–136.

Jones, E.G., and Hendry, S.H.C. (1984). Basket cells. In: *The Cerebral Cortex.* Vol. 1, Cellular Components of the Cerebral Cortex (A. Peters and E.G. Jones, eds.). Plenum Press, New York, pp. 309–336.

Jones, E.G., Hendry, S.H.C., and DeFelipe, J. (1987). GABA-peptide neurons of the primate cerebral cortex: A limited cell class. In: *The Cerebral Cortex.* Vol. 6, Further Aspects of Cortical Function, Including Hippocampus (E.G. Jones and A. Peters, eds.). Plenum Press, New York, pp. 237–266.

Jones, E.G., and Powell, T.P.S. (1970). An electron microscopic study of terminal degeneration in the neocortex of the cat. *Phil. Trans. Roy. Soc.* (London) ser. B 257, 29–43.

Jones, E.G., and Wise, S.P. (1977). Size, laminar and columnar distribution of efferent cells in the sensory-cortex of primates. *J. Comp. Neurol. 175,* 391–438.

Kaas, J.H., Nelson, R.J., Sur, M., Lin, C-S., and Merzenich, M. M. (1979). Multiple representations of the body within the primary somatosensory cortex of primates. *Science 204,* 521–523.

Kaitz, S.S., and Robertson, R.T. (1981). Thalamic connections with limbic cortex. II. Corticothalamic projections. *J. Comp. Neurol. 195,* 527–545.

Kaneko, T., and Misuno, N. (1988). Immunohistochemical study of glutaminase-containing neurons in the cerebral cortex and thalamus of the rat. *J. Comp. Neurol. 267,* 590–602.

Katz, L.C. (1987). Local circuitry of identified projections neurons in cat visual cortex brain slices. *J. Neurosci. 7,* 1223–1249.

Keller, A., and White, E.L. (1987). Synaptic organization of GABAergic neurons in the mouse SmI cortex. *J. Comp. Neurol. 262,* 1–12.

Keller, A., and White, E.L. (1989). Triads: A synaptic network component in the cerebral cortex. *Brain Res.,* in press.

Keller, A., White, E.L., and Cipolloni, P.B. (1985). The identification of thalamocortical axon terminals in barrels of mouse SmI cortex using immunohistochemistry of anterogradely transported lectin (*Phaseolus vulgaris*-leucoagglutinin). *Brain Res. 343,* 159–165.

Kelly, J.P., and Van Essen, D.C. (1974). Cell structure and function in the visual cortex of the cat. *J. Physiol. 238,* 515–547.

Kelly, J.P., and Wong, D. (1981). Laminar connections of the cat's auditory cortex *Brain Res. 212,* 1–15.

Kemper, T.L., and Galaburda, A.M. (1984). Principles of cytoarchitectonics. In: *The Cerebral Cortex.* Vol. 1, Cellular Components of the Cerebral Cortex (A. Peters and E.G. Jones, eds.). Plenum Press, New York, pp. 35–57.

Kennedy, C., Des Rosiers, H. M., Sakurada, O., Shinohara, M., Reivich, M., Jehle, J.W., and Sokoloff, L. (1976). Metabolic mapping of the primary visual system of the monkey by means of the autoradiographic [14C]deoxyglucose technique. *Proc. Natl. Acad. Sci. 73,* 4230–4234.

Kennedy, H., and Bullier, J. (1985). A double-labeling investigation of the afferent connectivity to cortical areas V1 and V2 of the macaque monkey. *J. Neurosci. 5,* 2815–2830.

Khachaturian, H., Lewis, M.E., Shafer, M.K.-H., and Watson, S.J. (1985). Anatomy of the CNS opioid systems. *TINS 8,* 111–119.

Kisvarday, S.F., Martin, K.A.C., Freund, T.F., Magloczky, Z.F., Whitteridge, D., and Somogyi, P. (1986c). Synaptic targets of HRP-filled layer III pyramidal cells in the cat striate cortex. *Exp. Brain Res. 64,* 541–552.

Kisvarday, S.F., Martin, K.A.C., Whitteridge, D., and Somogyi, P. (1985). Synaptic connections of intracellularly filled clutch cells: A type of small basket cell in the visual cortex of the cat. *J. Comp. Neurol. 241,* 111–137.

Kisvarday, Z.F., Adams, C.B.T., and Smith, A.D. (1986a). Synaptic connections of axo-axonic (chandelier) cells in human epileptic temporal cortex. *Neuroscience 19,* 1179–1186.

Kisvarday, Z.F., Cowey, A., and Somogyi, P. (1986b). Synaptic relationships of a type of GABA-immunoreactive neuron (clutch cell), spiny stellate cells and lateral geniculate nucleus afferents in layer IVC of the monkey striate cortex. *Neuroscience 19,* 741–761.

Kisvarday, Z.F., Martin, K.A.C., Friedlander, M.J., and Somogyi, P. (1987). Evidence for interlaminar inhibitory circuits in the striate cortex of the cat. *J. Comp. Neurol. 260,* 1–19.

Kita, H., and Oomura, Y. (1982). Evidence for a glycinergic corticolateral hypothalamic inhibitory pathway in the rat. *Brain Res. 235,* 131–136.

Kosar, E., and Hand, P. J. (1981). First somatosensory cortical columns and associated neuronal clusters of nucleus ventralis posterolateralis of the cat: An anatomical demonstration. *J. Comp. Neurol. 198,* 515–539.

Kosar, E., Waters, R. S., Tsukahara, N., and Asanuma, H. (1985). Anatomical and physiological properties of the projection from the sensory cortex to the motor cortex in normal cats: The difference between corticocortical and thalamocortical projections. *Brain Res. 345,* 68–78.

Krnjević, K. (1984). Neurotransmitters in cerebral cortex: A general account. In: *The Cerebral Cortex.* Vol. 2, Functional Properties of Cortical Cells (E.G. Jones and A. Peters, eds.). Plenum Press, New York, pp. 39–61.

Krnjević, K., Pumain, R., and Renaud, L. (1971). The mechanisms of excitation by acetylcholine in the cerebral cortex. *J. Pysiol. 215,* 247-268.

Kupfermann, I. (1985). Hemispheric asymmetries and the cortical localization of higher cognitive and affective functions. In: *Principles of Neuroscience* (E.R. Kandel and J.H. Schwartz, eds.). Elsevier, New York, pp. 673–687.

Lam, D.M., Su, Y.Y., and Watt, C.B. (1986). The self regulating synapse: A functional role for the existence of neuroactive substances. *Brain Res. 396,* 249–257.

Land, P.W., and Simons, D.J. (1985). Cytochrome oxidase staining in the rat SmI barrel cortex. *J. Comp. Neurol. 238,* 225–235.

Landry, P., and Deschênes, M. (1981). Intracortical arborizations and receptive fields of identified ventrobasal thalamocortical afferents to the primary somatic sensory cortex in the cat. *J. Comp. Neurol. 199,* 345–371.

Landry, P., Diadori, P., Leclerc, S., and Dykes, D.W. (1987). Morphological and electrophysiological characteristics of somatosensory thalamocortical axons studied with intra-axonal staining and recording in the cat. *Exp. Brain Res. 65,* 317–330.

LeVay, S. (1973). Synaptic patterns in the visual cortex of the cat and monkey. Electron microscopy of Golgi preparations. *J. Comp. Neurol. 150,* 53–86.

LeVay, S. (1986). Synaptic organization of claustral and geniculate afferents to the visual cortex of the cat. *J. Neurosci. 6,* 3564–3575.

Levin, B.E., Craik, R.L., and Hand, P.J. (1988). The role of norepinephrine in adult rat somatosensory (SmI) cortical metabolism and plasticity. *Brain Res. 443,* 261–271.

Lewis, W.B. (1878). On the comparative structure of the cortex cerebri. *Brain 1,* 79–86.

Lidov, H.G.W., Grzanna, R., and Molliver, M.E. (1980). The serotonin innervation of the cerebral cortex in the rat: An immunohistochemical analysis. *Neuroscience 5,* 207–227.

Lin, C.-S., Lu, S.M., and Schmechel, D.E. (1986). Glutamic acid decarboxylase and somatostatin immunoreactivities in rat visual cortex. *J. Comp. Neurol. 244,* 369–383.

Lin, C.S., Lu, S.M., and Yamawaki, R.M. (1987). Laminar and synaptic organization of terminals from the ventrobasal and posterior thalamic nuclei in rat barrel cortex. *Soci. Neurosci. Abst. 13,* part I, 248.

Lindvall, O., and Björklund, A. (1984). General organization of cortical monoamine systems. *Neurol. Neurobiol. 10,* 9–40.

Lorente de Nó, R. (1922). La corteza cerebral del raton. *Trab. Lab. Invest. Biol. Univ. Madrid 20,* 1–38.

Lorente de Nó, R. (1938). The physiology of the nervous system. In: *The Physiology of the Nervous System* (J.F. Fulton, ed.). Oxford Univ. Press, New York, pp. 291–329.

Lowel, S., Freeman, B., and Singer, W. (1987). Topographic organization of the orientation column system in large flat-mounts of cat visual cortex: A 2-deoxyglucose study. *J. Comp. Neurol. 255,* 401–415.

Lund, J.S. (1973). Organization of neurons in the visual cortex: Area 17, of the monkey (Macaca mulatta). *J. Comp. Neurol. 147,* 455–496.

Lund, J.S. (1984). Spiny stellate neurons. In: *The Cerebral Cortex.* Vol 1, Cellular Components of the Cerebral Cortex (A. Peters and E.G. Jones, eds.). Plenum Press, New York, pp. 255–308.

Lund, J.S. (1987). Local circuit neurons of macaque monkey striate cortex: I. Neurons of laminae 4C and 5A. *J. Comp. Neurol. 257,* 60–92.

Lund, J.S., and Boothe, R.G. (1975). Interlaminar connections and pyramidal neuron organization in the visual cortex, area 17, of the macaque monkey. *J. Comp. Neurol. 159,* 305–334.

Lund, J.S., Hendrickson, A.E., Ogren, M.P., and Tobin, E.A. (1981). Anatomical organization of primate visual cortex area VII. *J. Comp. Neurol. 202,* 19–45.

Lund, J.S., Henry, G.H., MacQueen, C.L., and Harvey, A.R. (1979). Anatomical organization of the primary visual cortex (area 17) of the cat. A comparison with area 17 of the Macaque monkey. *J. Comp. Neurol. 184,* 599–618.

Lund, J.S., and Lund, R.D. (1970). The termination of callosal fibers in the paravisual cortex of the rat. *Brain Res. 17,* 25–45.

Lund, J.S., Lund, R.D., Hendrickson, A.E., Bunt, A.H., and Fuchs, A.F. (1975). The origin of efferent pathways from the primary visual cortex, area

17, of the Macaque monkey as shown by retrograde transport of horseradish peroxidase. *J. Comp. Neurol. 164,* 287–304.

Marin-Padilla, M. (1969). Origin of the pericellular baskets of the pyramidal cells of the human motor cortex: A Golgi study. *Brain Res. 14,* 633–646.

Marin-Padilla, M. (1970). Prenatal and early postnatal ontogenesis of the human motor cortex: A Golgi study. II. The basket-pyramidal system. *Brain Res. 23,* 185–191.

Marin-Padilla, M. (1972). Double origin of the pericellular baskets of the pyramidal cells of the human motor cortex: A Golgi study. *Brain Res. 38,* 1–12.

Marin-Padilla, M. (1974). Three-dimensional reconstruction of the pericellular nests (baskets) of the motor (area 4) and visual (area 17) areas of the human cerebral cortex. A Golgi study. *Z. Anat. Entwickl.-Gesch. 144,* 123–135.

Marin-Padilla, M. (1984). Neurons of layer I: A developmental analysis. In: *The Cerebral Cortex.* Vol. 1, Cellular Components of the Cerebral Cortex (A. Peters and E.G. Jones, eds.). Plenum Press, New York, pp. 447–478.

Marin-Padilla, M. (1987). The chandelier cell of the human visual cortex: A Golgi study. *J. Comp. Neurol. 256,* 61–70.

Martin, K.A.C., Somogyi, P., and Whitteridge, D. (1983). Physiological and morphological properties of identified basket cells in the cat's visual cortex. *Exp. Brain Res. 50,* 193–200.

Martin, K.A.C., and Whitteridge, D. (1984). Form, function and intracortical projections of spiny neurones in the striate visual cortex of the cat. *J. Physiol. 353,* 463–504.

Maunsell, J.H.R., and Van Essen, D.C. (1983). The connections of the middle temporal visual area (MT) and their relationship to a cortical hierarchy in the macaque monkey. *J. Neurosci. 3,* 2563–2586.

McGeer, P.L., Eccles, J.C., and McGeer, E.G. (1987). *Molecular Neurobiology of the Mammalian Brain,* 2nd ed. Plenum Press, New York and London.

McGuire, B.A., Hornung, J.-P., Gilbert, C.D., and Wiesel, T.N. (1984). Patterns of synaptic input to layer 4 of cat striate cortex. *J. Neurosci. 4,* 3021–3033.

McMullen, N.T., and Glaser, E.M. (1982). Morphology and laminar distribution of nonpyramidal neurons in the auditory cortex of the rabbit. *J. Comp. Neurol. 208,* 85–106.

Meinecke, D.L., and Peters, A. (1987). GABA immunoreactive neurons in rat visual cortex. *J. Comp. Neurol. 261,* 388–404.

Merzenich, M.M., Kaas, J.H., Sur, M., and Lin, C.-S. (1978). Double representation of the body surface within cytoarchitectonic areas 3b and 2 in "SI" in owl monkey *(Aotus trivirgatus). J. Comp. Neurol. 181,* 41–74.

Meyer, G. (1982). Short-axon neurons with two axon-like processes in the visual cortex of the cat. A Golgi study. *Brain Res. 232,* 455–459.

Meyer, G., and Albus, K. (1981). Spiny stellates as cells of origin of association fibres from area 17 to area 18 in the cat's neocortex. *Brain Res. 210,* 335–341.

Meynert, T. (1869–1872). Von Gehirne der Säugethiere. In: *Handbuch der Lehre von den Geweben des Menschen und der Thiere* (Stricker, ed.). Cited in Brazier, 1978.

Minkowski, M. (1913). *Experimentalle Untersuchungen über die Beziehungen der Grosshirnrinde and Netzhaut zu den primären optischen Zentren, besonders zu Corpus Geniculatum externum, habilitationschr.* Zürich, p. 104. Cited in Putnam and Putnam, 1926.

Morrison, J.H., Foote, S.L., and Bloom, F. E. (1984). Laminar, regional, developmental, and functional specificity of monoaminergic innervation patterns of monkey cortex. *Neurol. Neurobiol. 10,* 61–75.

Mountcastle, V.B. (1957). Modality and topographic properties of single neurons of cat's somatic sensory cortex. *J. Neurophysiol. 20,* 408–434.

Mountcastle, V.B. (1978). An organizing principle for cerebral function: The unit module and the distributed system. In: *The Mindful Brain* (G.M. Edelman and V.B. Mountcastle, eds.). MIT Press, Cambridge, Mass., pp. 7–50.

Mountcastle, V.B. (1979). An organizing principle for cerebral function: The unit module and the distributed system. In: *The Neurosciences. Fourth Study Program* (F.O. Schmidt and F.G. Worden, eds.). MIT Press, Cambridge, Mass., pp. 21–42.

Müller-Paschinger, I.-B., Tombol, T., and Petsche, H. (1983). Chandelier neurons within the rabbits' cerebral cortex: A Golgi study. *Anat. Embryol. 166,* 149–159.

Naegele, J.R., Arimatsu, Y., Schwartz, P., and Barnstable, C.J. (1988). Selective staining of a subset of GABAergic neurons in cat visual cortex by monoclonal antibody VC1.1. *J. Neurosci. 8,* 79–89.

Nauta, W.J.H., and Karten, H.J. (1970). A general profile of the vertebrate brain, with sidelights on the ancestry of cerebral cortex. In: *The Neurosciences: Second Study Program* (F.O. Schmitt, ed.). Rockefeller Univ. Press, New York, pp. 7–25.

Olavarria, J., and Van Sluyters, R.C. (1985). Organization and postnatal development of callosal connections in the visual cortex of the rat. *J. Comp. Neurol. 239,* 1–26.

O'Leary, J.L. (1941). Structure of the area striate of the cat. *J. Comp. Neurol. 75,* 131–163.

Ottersen, O.P., Fisher, B.O., and Storm-Mathisen, J. (1983). Retrograde transport of D-[3H] aspartate in thalamocortical neurons. *Neurosci. Lett. 42,* 19–24.

Pandya, D.N., and Yeterian, E.H. (1985). Architecture and connections of cortical association areas. In: *The Cerebral Cortex.* Vol. 4, Association and Auditory Cortices (A. Peters and E.G. Jones, eds.). Plenum Press, New York, pp. 3–61.

Papadopoulos, G.C., Parnavelas, J.G., and Cavanagh, M.E. (1987). Extensive co-existence of neuropeptides in the rat visual cortex. *Brain Res. 420,* 95–99.

Pappas, C.L., and Strick, P.L. (1981). Physiological demonstration of multiple representation in the forelimb region of the cat motor cortex. *J. Comp. Neurol. 200,* 481–490.

Parnavelas, J.G. (1984). Physiological properties of identified neurons. In: *Cerebral Cortex.* Vol. 2, (E.G. Jones and A. Peters, eds.). Plenum Press, New York, pp. 205–239.

Parnavelas, J.G., Burne, R.A., and Lin, C.-S. (1983). Distribution and morphology of functionally identified neurons in the visual cortex of the rat. *Brain Res. 261,* 21–29.

Parnavelas, J.G., Kelly, W., Franke, E., and Eckenstein, F. (1986). Cholinergic neurons and fibres in the rat visual cortex. *J. Neurocytol. 15,* 329–336.

Parnavelas, J.G., Sullivan, K., Lieberman, A.R., and Webster, K.E. (1977). Neurons and their synaptic organization in the visual cortex of the cat. *Cell Tiss. Res. 183,* 499–517.

Penfield, W., and Rasmussen, T. (1950). *The Cerebral Cortex of Man: A Clinical Study of Localization of Function.* Macmillan, New York.

Penny, G.R., Arsharpour, S., and Kitai, S.T. (1986). Substance P-immunoreactive neurons in the neocortex of the rat: A subset of the glutamic acid decarboxylase-immunoreactive neurons. *Neurosci. Lett. 65,* 53–59.

Peters, A. (1984). Bipolar cells. In: *The Cerebral Cortex.* Vol. 1, Cellular Components of the Cerebral Cortex (A. Peters and E.G. Jones, eds.). Plenum Press, New York, pp. 381–407.

Peters, A. (1987a). The bipolar cell in rat visual cortex. *Society for Neuroscience Abstracts 13,* p. 359.

Peters, A. (1987b). Number of neurons and synapses in primary visual cortex. In: *The Cerebral Cortex.* Vol. 6, (E.G. Jones and A. Peters, eds.). Plenum Press, New York, pp. 267–294.

Peters, A., and Fairén, A. (1978). Smooth and sparsely-spined stellate cells in the visual cortex of the rat: A study using a combined Golgi-electron microscope technique. *J. Comp. Neurol. 181,* 129–172.

Peters, A., and Feldman, M.L. (1976). The projection of the lateral geniculate nucleus to area 17 of rat cerebral cortex. I. General Description. *J. Neurocytol. 5,* 63–84.

Peters, A., Feldman, M., and Saldanha, J. (1976). The projection of the lateral geniculate nucleus to area 17 of the rat cerebral cortex. II. Terminations upon neuronal perikarya and dendritic shafts. *J. Neurocytol. 5,* 85–107.

Peters, A., and Harriman, K.M. (1988). Enigmatic bipolar cell of rat visual cortex. *J. Comp. Neurol. 267,* 409–432.

Peters, A., and Jones, E. G. (1984). Classification of cortical neurons. In: *The Cerebral Cortex.* Vol. 1, Cellular Components of the Cerebral Cortex (A. Peters and E.G. Jones, eds.). Plenum Press, New York, pp. 107–121.

Peters, A., and Kaiserman-Abramof, I. (1970). The small pyramidal neuron of the rat cerebral cortex: The perikaryon, dendrites and spines. *Am. J. Anat. 127,* 321–356.

Peters, A., and Kara, D.A. (1985). The neuronal composition of area 17 of rat visual cortex. I. The pyramidal cells. *J. Comp. Neurol. 234,* 218–241.

Peters, A., and Kara, D.A. (1987). The neuronal composition of area 17 of rat visual cortex. IV. The organization of pyramidal cells. *J. Comp. Neurol. 260,* 573–590.

Peters, A., Kara, D.A., and Harriman, K. M. (1985). The neuronal composition of area 17 of rat visual cortex. III. Numerical considerations. *J. Comp. Neurol. 238,* 263–274.

Peters, A., and Kimerer, L.M. (1981). Bipolar neurons in rat visual cortex: A combined Golgi-electron microscope study. *J. Neurocytol. 10,* 921–946.

Peters, A., Meinecke, D.L., and Karamanlidis, A.N. (1987c). Vasoactive intestinal polypeptide immunoreactive neurons in the primary visual cortex of the cat. *J. Neurocytol. 16,* 23–38.

Peters, A., Palay, S.L., and Webster, H. deF. (1976). *The Fine Structure of the Nervous System: The Neurons and Supporting Cells* (A. Peters, ed.). W.B. Saunders Co., Philadelphia.

Peters, A., and Proskauer, C.C., (1980). Synaptic relations between a multipolar stellate cell and a pyramidal neuron in the rat visual cortex. A combined Golgi-electron microscope study. *J. Neurocytol. 9,* 163–183.

Peters, A., Proskauer, C.C., Feldman, M.L., and Kimerer, L. (1979). The projection of the lateral geniculate nucleus to area 17 of the rat cerebral cortex. V. Degenerating axon terminals synapsing with Golgi impregnated neurons. *J. Neurocytol. 8,* 331–357.

Peters, A., Proskauer, C.C. and Ribak, C.E. (1982). Chandelier cells in rat visual cortex. *J. Comp. Neurol. 206,* 397–416.

Peters, A., and Regidor, J. (1981). A reassessment of the forms of nonpyramidal neurons in area 17 of cat visual cortex. *J. Comp. Neurol. 203,* 685–716.

Peters, A., and Saint Marie, R.L. (1984). Smooth and sparsley spinous nonpyramidal cells forming local axonal plexuses. In: *The Cerebral Cortex.* Vol. 1, Cellular Components of the Cerebral Cortex (A. Peters and E.G. Jones, eds.). Plenum Press, New York, pp. 419–445.

Peters, A., and Saldanha, J. (1976). The projection of the lateral geniculate nucleus to area 17 of the rat cerebral cortex. III. Layer VI. *Brain Res. 105,* 533–537.

Peters, A., and Walsh, T.M. (1972). A study of the organization of apical dendrites in the somatic sensory cortex of the rat. *J. Comp. Neurol. 144,* 253–268.

Peters, A., White, E.L., and Fairén, A. (1977). Synapses between identified neuronal elements. An electron microscopic demonstration of degenerating axon terminals synapsing with Golgi impregnated neurons. *Neurosci. Lett. 6,* 171–175.

Phillipson, O.T., Kilpatrick, I.C., and Jones, M.W. (1987). Dopaminergic innervation of the primary visual cortex in the rat, and some correlations with human cortex. *Brain Res. Bull. 18,* 621–633.

Polyak, S. (1933). A contribution to the cerebral representation of the retina. *J. Comp. Neurol. 57,* 541–567.

Polyak, S. (1957). *The Vertebrate Visual System* (H. Kluver, ed.). Univ. of Chicago Press, Chicago.

Porter, L., and White, E.L. (1984). Termination of callosal afferents onto identified callosal projection neurons. *Neurosci. Lett. 47,* 37–40.

Porter, L., and White, E.L. (1986). Synaptic connections of callosal projection neurons in the vibrissal region of mouse primary motor cortex: An electron microscopic/horseradish peroxidase study. *J. Comp. Neurol. 253,* 303–314.

Putnam, T.J., and Putnam, I.K. (1926). Studies on the central visual system. I. The anatomic projection of the retinal quadrants on the striate cortex of the rabbit. *Arch. Neurol. Neurosurg. Psychiat. 16,* 1–20.

Rall, W., and Segev, I. (1987). Functional possibilities for synapses on dendrites and dendritic spines. In: *Synaptic Function* (G.M. Edelman, W.F. Gall, and W.M. Cowan, eds.). Wiley, New York, pp. 605–636.

Ribak, C.E. (1978). Aspinous and sparsely-spinous stellate neurons in the visual cortex of rats contain glutamic acid decarboxylase. *J. Neurocytol. 7,* 461–478.

Robinson, C.J., and Burton, H. (1980a). Somatotopic organization in the second somatosensory area of *M. fascicularis. J. Comp. Neurol. 192,* 43–68.

Robinson, C.J., and Burton, H. (1980b). Somatic submodality distribution within the second somatosensory (SII), 7b, retroinsular, postauditory, and granular insular cortical areas of *M. fascicularis. J. Comp. Neurol. 192,* 93–108.

Rockel, A.J., Hiorns, R.W., and Powell, T.P.S. (1980). The basic uniformity in structure of the neocortex. *Brain 103,* 221–244.

Rockland, K.S., and Lund, J.S. (1983). Intrinsic laminar lattice connections in primate visual cortex. *J. Comp. Neurol. 216,* 303–318.

Rockland, K.S., Lund, J.S., and Humphrey, A.L. (1982). Anatomical banding of intrinsic connections in striate cortex of tree shrews *(tupaia glis). J. Comp. Neurol. 209,* 41–58.

Romagno, M.A., and Maciewicz, R.J. (1975). Peroxidase labelling of motor cortex neurons projecting to ventrolateral nucleus in the cat. *Brain Res. 83,* 469–473.

Rose, J.E., and Woolsey, C.N. (1949). Organization of the mammalian thalamus and its relationship to the cerebral cortex. *Electroenceph. Clin. Neurophysiol., 7,* 391–404.

Rose, J.E., and Woolsey, C.N. (1958). Cortical connections and functional organization of the thalamic auditory system of the cat. In: *Biological and Biochemical Bases of Behavior* (H.F. Harlow and C.N. Woolsey, eds.). Univ. of Wisconsin Press, Madison, pp. 127–150.

Rose, M. (1929). Cytoarchitektonischer Atlas der Grosshirnde der Maus. *Journal für Psychologie und Neurologie,* 40, 1–51. Cited in Braak (1984).

Rose, M. (1935). Cytoarchitektonik und Myeloarchitektonik der Grosshirnrinde. In: *Handbuch der Neurologie,* vol. 1 (O. Bumke and O. Förster, eds.). Springer, Berlin, pp. 588–778.

Rosenquist, A.C. (1985). Connections of visual cortical areas in the cat. In: *The Cerebral Cortex.* Vol. 3, Visual Cortex (A. Peters and E.G. Jones, eds.). Plenum Press, New York, pp. 81–117.

Saint Marie, R.L., and Peters, A. (1985). The morphology and synaptic connections of spiny stellate neurons in monkey visual cortex (area 17): A Golgi-electron microscopic study. *J. Comp. Neurol. 233,* 213–235.

Sato, H., Hata, Y., Masui, H., and Tsumoto, T. (1987). A functional role of cholinergic innervation to neurons in the cat striate cortex. *J. Neurophysiol. 58,* 765–780.

Schmechel, D. E., Vickrey, B.G., Fitzpatrick, D., and Elde, R.P. (1984). GABAergic neurons of mammalian cerebral cortex: Widespread subclass defined by somatostatic content. *Neurosci. Lett. 47,* 227–232.

Schober, W., Hedlich, A., Müller, L., and Vrensen, G. (1983a). Geniculocortical synapses involving apical dendrites of layer V pyramidal cells in area

17 of the albino rat visual cortex. A combined Golgi/EM study. *J. Hirnforsch.* *24,* 113–121.

Schober, W., Hedlich, A., Vrensen, G., and Müller, L. (1983b). Termination of geniculocortical afferents on bipolar neurons in area 17 of the albino rat. A Golgi/EM study. *J. Hirnforsch.* *24,* 473–477.

Schüz, A., and Münster, A. (1985). Synaptic density on the axonal tree of a pyramidal cell in the cortex of the mouse. *Neuroscience* *15,* 33–39.

Sholl, D.A. (1953). Dendritic organization in the visual and motor cortices of the cat. *J. Anat.* *87,* 387–407.

Sholl, D.A. (1967). *The Organization of the Cerebral Cortex* (facsimile of the 1956 ed.). Hafner Publishing Co., London, 1967.

Sillito, A.M., and Murphy, P.C. (1987). The cholinergic modulation of cortical function. In: *The Cerebral Cortex.* Vol. 6, Further Aspects of Cortical Function, Including Hippocampus (E.G. Jones and A. Peters, eds.). Plenum Press, New York, pp. 161–187.

Simons, D.J. (1978). Response properties of vibrissa units in rat SI somatosensory neocortex. *J. Neurophysiol.* *41,* 798–820.

Simons, D.J. (1985). Temporal and spatial integration in the rat SI vibrissa cortex. *J. Neurophysiol.* *54,* 615–635.

Simons, D.J., and Woolsey, T.A. (1979). Functional organization in mouse barrel cortex. *Brain Res.* *165,* 327–332.

Simons, D.J., and Woolsey, T.A. (1984). Morphology of Golgi-Cox-impregnated barrel neurons in rat SmI cortex. *J. Comp. Neurol.* *230,* 119–132.

Sloper, J.J. (1973). An electron microscope study of the termination of afferent connections to the primate motor cortex. *J. Neurocytol.* *2,* 361–368.

Sloper, J.J., Hiorns, R.W., and Powell, T.P.S. (1979). A qualitative and quantitative electron microscopic study of the neurons in the primate motor and somatic sensory cortices. *Phil. Trans. Roy. Soc.* (London) *285,* 141–171.

Sloper, J.J., and Powell, T.P.S. (1979). An experimental electron microscopic study of afferent connections to the primate motor and somatic sensory cortices. *Phil. Trans. Roy. Soc.* (London) ser. B. 285, 199–225.

Smith, L.M., Ebner, F.F., and Colonnier, M. (1980). The thalamocortical projection in pseudemys turtles: A quantitative electron microscopic study. *J. Comp. Neurol.* *190,* 445–461.

Somogyi, P. (1977). A specific "axo-axonal" interneuron in the visual cortex of the rat. *Brain Res.* *136,* 345–350.

Somogyi, P. (1978). The study of Golgi stained cells and of experimental degeneration under the electron microscope: A direct method for the identification in the visual cortex of three successive links in a neuron chain. *Neuroscience* *3,* 167–180.

Somogyi, P. (1979). An interneurone making synapses specifically on the axon initial segment of pyramidal cells in the cerebral cortex. *J. Physiol.* *296,* 18–19P.

Somogyi, P., and Cowey, A. (1981). Combined Golgi and electron microscopic study on the synapses formed by double bouquet cells in the visual cortex of the cat and monkey. *J. Comp. Neurol.* *195,* 547–566.

Somogyi, P., and Cowey, A. (1984). Double bouquet cells, In: *The Cerebral Cortex*. Vol. 1, Cellular Components of the Cerebral Cortex (A. Peters and E.G. Jones, eds.). Plenum Press, New York, pp. 337–360.

Somogyi, P., Cowey, A., Halasz, N., and Freund, T.F. (1981). Vertical organization of neurons accumulating 3H-GABA in the visual cortex of the rhesus monkey. *Neuroscience 7,* 2577–2609.

Somogyi, P., Freund, T.F., and Cowey, A. (1982). The axo-axonic interneuron in the cerebral cortex of the rat, cat and monkey. *Neuroscience. 7,* 2577–2609.

Somogyi, P., Freund, T.F., Hodgson, A.J., Somogyi, J., Beroukas, D., and Chubb, I.W. (1985). Identified axo-axonic cells are immunoreactive for GABA in the hippocampus and visual cortex of the cat. *Brain Res. 332,* 143–149.

Somogyi, P., Hodgson, A.J., and Smith, A.D. (1979). An approach to tracing neuron networks in the cerebral cortex and basal ganglia. Combination of Golgi staining, retrograde transport of horseradish peroxidase and anterograde degeneration of synaptic boutons in the same material. *Neuroscience 4,* 1805–1852.

Somogyi, P., Hodgson, A.J., Smith, A.D., Nunzi, M.G., Gorio, A., and Wu, J.Y. (1984). Different populations of gabaergic neurons in the visual cortex and hippocampus of cat contain somatostatin- or cholecystokinin-immunoreactive material. *J. Neurosci. 4,* 2590–2603.

Somogyi, P., Kisvarday, Z.F., Martin, K.A.C., and Whitteridge, D. (1983a). Synaptic connections of morphologically identified and physiologically characterized large basket cells in the striate cortex of the cat. *Neuroscience 10,* 261–294.

Somogyi, P., Nunzi, M.G., Gorio, A., and Smith, A.D. (1983b). A new type of specific interneuron in the monkey hippocampus forming synapses exclusively with the axon initial segments of pyramidal cell. *Brain Res. 259,* 137–142.

Somogyi, P., and Soltész, I. (1986). Immunogold demonstration of GABA in synaptic terminals of intracellularly recorded, horseradish peroxidase-filled basket cells and clutch cells in the cat's visual cortex. *Neuroscience 19,* 1051–1065.

Sretavan, D., and Dykes, R.W. (1983). The organization of two cutaneous submodalities in the forearm region of area 3b of cat somatosensory cortex. *J. Comp. Neurol. 213,* 381–398.

Streit, P. (1984). Glutamate and aspartate as transmitter candidates for systems of the cerebral cortex. In: *The Cerebral Cortex*. Vol. 2, (E.G. Jones and A. Peters, eds). Plenum Press, New York, pp. 119–143.

Streit, P., and Cuenod, M. (1979). Transmitter specificity and connectivity revealed by differential retrograde labeling of neuroal pathways. *Neurosci. Lett.*, Suppl. 3, 340.

Strick, P.L., and Preston, J.B. (1978). Multiple representations in the primate motor cortex. *Brain Res. 154,* 366–370.

Strick, P.L., and Sterling, P. (1974). Synaptic termination of afferents from the ventrolateral nucleus of the thalamus in the cat motor cortex. A light and electron microscope study. *J. Comp. Neurol. 153,* 77–106.

Swadlow, H.A., Weyand, T.G., and Waxman, S.G. (1978). The cells of origin of the corpus callosum in rabbit visual cortex. *Brain Res. 156,* 129–134.

Szentágothai, J. (1965). In: *Progress in Brain Research,* vol. 14, The Use of Degeneration in the Investigation of Short Neuronal Connections (M. Singer and J.P. Shade eds.). Elsevier, Amsterdam, pp. 1–32.

Szentágothai, J. (1975). The "module-concept" in cerebral cortex architecture. *Brain Res. 95,* 475–496.

Talbot, S.A., and Marshall, W.H. (1941). Physiological studies on neuronal mechanisms of visual localization and discrimination. *Am J. Ophth. 24,* 1255–1263.

Tanji, J., and Wise, S.P. (1981). Submodality distribution in sensorimotor cortex of the unanesthetized monkey. *J. Neurophysiol. 45,* 467–481.

Taylor, D.A., and Stone, T.W. (1981). Neurotransmodulatory control of cerebral cortical neuron activity. In: *The Organization of the Cerebral Cortex* (G. Adelman, S.G. Dennis, F.O. Schmitt, and F.G. Worden, eds.). MIT Press, Cambridge, Mass., pp. 347–357.

Tigges, J., and Tigges, M. (1982). Principles of axonal collateralization of lamina II-III pyramids in area 17 of squirrel monkey: A quantitative Golgi study. *Neurosci. Lett. 29,* 99–104.

Tigges, M., and Tigges. J. (1979). Types of degenerating geniculocortical axon terminals and their contribution to layer IV of area 17 in the squirrel monkey (Samiri). *Cell Tissue Res. 196,* 471–486.

Tömbol, T. (1978). Comparative data on the Golgi architecture of interneurons of different cortical areas in cat and rabbit. In: *Architectonics of the Cerebral Cortex.* (M.A.B. Brazier and H. Petsche eds.). Raven Press, New York, pp. 59–76.

Tootell, R.B.H., Silverman, M. S., Switkes, E., and DeValois, R.L. (1982). Deoxyglucose analysis of retinotopic organization in primate striate cortex. *Science 218,* 902–904.

Tsumoto, T., Masui, H., and Sato, H. (1986). Excitatory amino acid transmitters in neuronal circuits of the cat visual cortex. *J. Neurophysiol. 55,* 469–483.

Valverde, F. (1978). The organization of area 18 in the monkey: A Golgi study. *Anat. Embryol. 154,* 305–334.

Valverde, F. (1983). A comparative approach to neocortical organization based on the study of the brain of the hedgehog (Erinaceus europaeus). In: *Ramon y Cajal's Contribution to the Neurosciences* (S. Grisiola, C. Guerri, F. Samson, S. Norton, and F. Reinoso-Suarez, eds.). Elsevier, Amsterdam, pp. 149–170.

Valverde, F. (1984). The organizing principles of the primary visual cortex in the monkey. In: *The Cerebral Cortex.* Vol. 1, Cellular Components of the Cerebral Cortex (A. Peters and E.G. Jones, eds.). Plenum Press, New York, pp. 207–257.

Van der Loos, H. (1967). The history of the neuron. In: *The Neuron* (H. Hyden, ed.). Elsevier, Amsterdam-London-New York, pp. 1–47.

Van Essen, D.C. (1985). Functional organization of primate visual cortex. In: *The Cerebral Cortex.* Vol. 3, Visual Cortex (A. Peters and E.G. Jones, eds.). Plenum Press, New York, pp. 259–329.

Van Essen, D.C., and Kelly, J. (1973). Correlation of cell shape and function in the visual cortex of the cat. *Nature 241,* 403–405.

Van Essen, D.C., and Maunsell, J.H.R. (1980). Two-dimensional maps of the cerebral cortex. *J. Comp. Neurol. 191,* 255–281.

Van Essen, D.C., and Maunsell, J.H.R. (1983). Hierarchical organization and functional streams in the visual cortex. *T.I.N.S.,* Sept., 370–375.

Vanegas, H., Hollander, H., and Distel, H. (1977). Early stages of uptake and transport of horseradish peroxidase by cortical structures, and its use for the study of local neurons and their processes. *J. Comp. Neurol. 177,* 193–212.

Vogt, B.A., and Peters, A. (1981). Form and distribution of neurons in rat cingulate cortex: Area 32, 24, and 29. *J. Comp. Neurol. 195,* 603–625.

von Economo, C., and Koskinas, G.N. (1925). *Die Cytoarchitektonic der Hirnrinde des erwachsenen Menschen.* Springer, Berlin. Cited in Braak (1984).

Walker, A.E. (1938). *The Primate Thalamus.* Univ. of Chicago Press, Chicago.

Walshe, F.M.R. (1942). The giant pyramidal cells of Betz, the motor cortex and the pyramidal tract: A critical review. *Brain 65,* 409–461.

Wamsley, J.K. (1984). Autoradiographic localization of receptor sites in the cerebral cortex. In: *The Cerebral Cortex,* Vol. 2, Functional Properties of Cortical Cells (E.G. Jones, and A. Peters, eds.). Plenum Press, New York, pp. 173–202.

Waterhouse, B.D., and Woodward, D.J. (1980). Interaction of norepinephrine with cerebrocortical activity evoked by stimulation of somatosensory afferent pathways in the rat. *Exp. Neurol. 67,* 11–34.

Welker, W.I. (1976). Mapping the brain. Historical trends in functional localization. *Brain Behav. Evol. 13,* 327–343.

White, E.L. (1976). Ultrastructure and synaptic connections in barrels of mouse SI cortex. *Brain Res. 105,* 229–251.

White, E.L. (1978). Identified neurons in mouse SmI cortex which are postsynaptic to thalamocortical axon terminals: A combined Golgi-electron microscopic and degeneration study. *J. Comp. Neurol. 181,* 627–662.

White, E.L. (1979). Thalamocortical synaptic relations: A review with emphasis on the projection of specific thalamic nuclei to the primary sensory areas of the neocortex. *Brain Res. Rev. 1,* 275–311.

White, E.L. (1981). Thalamocortical synaptic relations. In: *The Organization of the Cerebral Cortex* (G. Aldelman, S.G. Dennis, F.O. Schmitt, and F.G. Worden, eds.). MIT Press, Cambridge, Mass., pp. 153–161.

White, E.L. (1986). Terminations of thalamic afferents. In: *The Cerebral Cortex.* Vol. 5, Sensory-motor areas and Aspects of Cortical Connectivity (E.G. Jones and A. Peters eds.). Plenum Publishing Press, New York, pp. 271–289.

White, E.L. (1987). Thalamocortical interactions. In: *Encyclopedia of Neuroscience,* vol. II (G. Adelman, ed.). Birkhaüser Press, Boston, pp. 1202–1204.

White, E.L., Benshalom, G., and Hersch, S.M. (1984). Thalamocortical and other synapses of non-spiny multipolar cells in mouse SmI cortex. *J. Comp. Neurol. 229,* 311–320.

White, E.L., and DeAmicis, R. (1977). Afferent and efferent projections of the region in mouse SmI cortex which contains the posteromedial barrel subfield. *J. Comp. Neurol. 175,* 455–481.

White, E.L., and Hersch, S.M. (1981). Thalamocortical synapses of pyramidal cells which projects from SmI to MsI cortex in the mouse. *J. Comp. Neurol. 198,* 167–181.

White, E.L., and Hersch, S.M. (1982). A quantitative study of thalamocortical and other synapses involving the apical dendrites of corticothalamic projection cells in mouse SmI cortex. *J. Neurocytol. 11,* 137–157.

White, E.L., Hersch, S.M., and Belford, G.R. (1982). Quantitative studies of thalamocortical synapses with labeled pyramidal cells in mouse SmI cortex. *Soc. Neurosci. VIII,* 853.

White, E.L., Hersch, S.M., and Rock, M.P. (1980). Synaptic sequences in mouse SmI cortex involving pyramidal cells labeled by retrograde filling with horseradish peroxidase. *Neurosci. Lett. 19,* 149–154.

White, E.L., and Keller, A. (1987). Intrinsic circuitry involving the local axonal collaterals of corticothalamic projection cells in mouse SmI cortex. *J. Comp. Neurol. 262,* 13–26.

White, E.L., and Rock, M.P. (1979). Distribution of thalamic input to different dendrites of a spiny stellate cell. *Neurosci. Lett. 15,* 115–119.

White, E.L., and Rock, M.P. (1980). Three dimensional aspects and synaptic relationships of a Golgi impregnated spiny stellate cell reconstructed from serial thin sections. *J. Neurocytol. 9,* 615–636.

White, E.L., and Rock, M.P. (1981). A comparison of thalamocortical and other synaptic inputs to dendrites of two non-spiny neurons in a single barrel of mouse SmI cortex. *J. Comp. Neurol. 195,* 265–277.

Wiesel, T.N., Hubel, D. H., and Lam, D.M.K. (1974). Autoradiographic demonstration of ocular-dominance columns in the monkey striate cortex by means of transneuronal transport. *Brain Res. 79,* 273–279.

Winer, J.A. (1984). The pyramidal neurons in layer III of cat primary auditory cortex (AI). *J. Comp. Neurol. 229,* 476–496.

Winfield, D.A., Brooke, R.N.L., Sloper, J.J., and Powell, T.P.S. (1981). A combined Golgi-electron microscopic study of the synapses made by the proximal axon and recurrent collaterals of a pyramidal cell in the somatic sensory cortex of the monkey. *Neuroscience 6,* 1217–1230.

Winfield, D.A., Gatter, K.C., and Powell, T.P.S. (1980). An electron microscopic study of the types and proportions of neurons in the cortex of the motor and visual areas of the cat and rat. *Brain 103,* 245–258.

Winfield, D.A., and Powell, T.P.S. (1983). Laminar cell counts and geniculocortical boutons in area 17 of cat and monkey. *Brain Res. 277,* 223–229.

Winguth, S.D., and Winer, J.A. (1986). Corticocortical connections of cat primary auditory cortex (AI): Laminar organization and identification of supragranular neurons projecting to area AII. *J. Comp. Neurol. 248,* 36–56.

Wise, S.P., and Jones, E.G. (1977). Cells of origin and terminal distribution of descending projections of the rat somatic sensory cortex. *J. Comp. Neurol. 175,* 129–158.

Wong-Riley, M.T.T., and Welt, C. (1980). Histochemical changes in cytochrome oxidase of cortical barrels after vibrissal removal in neonatal and adult mice. *Proc. Natl. Acad. Sci.* 77, 2333–2337.

Woolsey, C.N. (1947). Patterns of sensory representation in the cerebral cortex. *Fed. Proc.* 6, 437–441.

Woolsey, C.N. (1958). Organization of somatic sensory and motor areas of the cerebral cortex. In: *Biological and Biochemical Bases of Behavior* (H.F. Harlow, and C.N. Woolsey, eds.). Univ. of Wisconsin Press, Madison.

Woolsey, C.N. (1964). Cortical localization as defined by evoked potential and electrical stimulation studies. In: *Cerebral Localization and Organization* (G. Schaltenbrand and C.N. Woolsey, eds.). Univ. of Wisconsin Press, Madison. pp. 17–26.

Woolsey, C.N., and Fairman, D. (1946). Contralateral, ipsilateral and bilateral representation of cutaneous receptors in somatic areas I and II of the cerebral cortex of pig, sheep and other mammals. *Surgery* 19, 684–702,

Woolsey, C.N., Marshall, W.H., and Bard, P. (1942). Representation of cutaneous tactile sensibility in the cerebral cortex of the monkey as indicated by evoked potentials. *Bull. Johns Hopkins. Hosp.* 70, 399–441.

Woolsey, C.N., and Walzl, E.M. (1942). Topical projection of nerve fibers from local regions of the cochlea to the cerebral cortex of the cat. *Bull. Johns Hopkins Hosp.* 71, 315–344.

Woolsey, T.A. (1967). Somatosensory, auditory and visual cortical areas of the mouse. *Johns Hopkins Med. J.* 121, 91–112.

Woolsey, T.A., Dierker, M.K., and Wann, D.F. (1975). Mouse SmI cortex: Qualitative and quantitative classification of Golgi-impregnated barrel neurons. *Proc. Nat. Acad. Sci.* 72, 2165–2169.

Woolsey, T.A., and Dierker, M.L. (1978). Computer assisted recording of anatomical data. In: *Neuroanatomical Research Techniques* (R.T. Robertson, ed.). Academic Press, New York, pp. 47–85.

Woolsey, T.A., and Van der Loos, H. (1970). The structural organization of layer IV in the somatosensory (SI) region of mouse cerebral cortex. *Brain Res.* 17, 205–242.

Part II Functional Aspects of Cortical Circuitry

ASAF KELLER

Part I presented the anatomical basis for cortical circuitry: the various neuronal types and the synaptic connections between them. The aim of Part II is twofold: first, to provide data from functional studies that reveal synaptic interactions between identified neurons, in order to extend the anatomical data on identified circuits; and second, to discuss the mechanisms by which identified cortical circuits may determine the functional properties of cortical neurons. Data related to the existence and operation of functional columns in the cerebral cortex will be presented at the outset, because of the prominent role played by functional columns in cortical circuitry.

Functional Properties of Cortical Neurons 4

Functional Columns

Physiological Identification of Functional Columns

A major turning point in the use of electrophysiology to study the cerebral cortex occurred in the late 1930s with the introduction of surface electrodes to record the activity of neurons in a large volume of cortex. In a series of pioneering studies, Woolsey et al. (e.g., 1942; and see Woolsey, 1964) placed wire electrodes on the pial surface of cat and monkey somatosensory cortex and recorded the electrical responses of neurons following stimulation of peripheral tissue. The results of these studies showed that cortical neurons are activated by discrete peripheral stimuli and that coherent maps in the brain can be produced by relating a region on the body surface to the point of maximal evoked activity in the cortex. Similar evoked-potential maps were produced for other cortical areas and in other species, including the somatosensory cortex of the human (Penfield and Boldrey, 1937), the gustatory cortex of the rat (Benjamin and Pfaffamann, 1955), and the visual cortex of the monkey (Marshall and Talbot, 1942) and human (Blakemore and Campbell, 1969).

A limitation of the evoked potential method is that it records the activity of many elements simultaneously, and so this approach cannot be used to study the responses of single neurons. This limitation was surmounted by the introduction of microelectrodes, which, since the 1940s, have provided neurophysiologists with a valuable tool by which the activity of individual neurons can be studied. At first, microelectrodes were placed extracellularly (e.g., Adrian and Morruzi, 1939), but by the early 1950s, methods were introduced to record intracellularly from neurons (e.g., Amassian, 1953; Tasaki et al., 1954).

Extracellular recordings from microelectrodes that are placed in close proximity to neuronal somata provide information on the responses of single cells to various stimuli without actually penetrating the neurons from which the recordings are taken. Using this approach, Mountcastle (1957) examined the responses to somatosensory stimuli of single neurons, at different cortical depths. He observed that neurons in all layers of the somatosensory cortex of the cat (Mountcastle, 1957) and monkey (Powell and Mountcastle, 1959) can be activated by mechanical stimulation of peripheral tissue and that a single neuron responds only to a specific somatosensory submodality, such as movement of hairs, pressure upon the skin, or mechanical deformation of deep tissues. Furthermore, individual neurons were activated only by stimuli applied to a particular area of the skin, and all neurons encountered by the electrode in vertical penetrations of the cortex responded to the same somatosensory submodality. In effect, Mountcastle's pioneering studies demonstrated that cortical neurons sharing similar receptive field properties are arranged within vertical columns.

These findings are consistent with the hypothesis set forth by Lorente de Nó (1938) that the elementary pattern of organization in the cerebral cortex is a vertically oriented column of neurons. Vertically oriented functional columns related to specific response properties have been demonstrated since in many cortical areas of various species (for a review, see Mountcastle, 1979). For instance, in the visual cortex, neurons that respond preferentially to stimuli presented to one eye or to stimuli having a particular orientation, wavelength, or direction of movement are grouped together in independently organized, vertically oriented multicellular arrays (Berman et al., 1987; Hubel and Wiesel, 1962, 1977; LeVay et al., 1987; Michael, 1978). Other studies have shown that, in the cat and monkey primary somatosensory cortices, rapidly adapting neurons, which respond only at the onset and offset of skin indentation, and slowly adapting neurons, which respond throughout the stimulus, are situated within separate slablike columns of cortex (Dykes et al., 1980; Dykes and Gabor, 1981; Sur et al., 1981, 1984).

Columns of cells having similar receptive field properties have also been identified in the primary (Favorov et al., 1987) and secondary (Alloway and Burton, 1985) somatosensory cortices of the cat. In the primary auditory cortex of the cat are columns of cells that respond to the same or to similar frequencies or to auditory stimuli from one (monaural) or both (binaural) ears (e.g., Abeles and Goldstein, 1970; Imig and Adrián, 1977). Binaural neurons that summate the input from both ears and those that suppress the input from one of the ears are arrayed within separate columns (Imig and Adrián, 1977). In the motor cortex, stimulation of discrete columns of neurons causes individual groups of muscles to contract (Asanuma and Rosén, 1972; Asanuma, 1975). Furthermore, neurons within a single efferent column in motor cortex receive proprioceptive input from the same joints or muscles that are activated by cells in that column, as well as input from skin superficial to the target muscles (Rosén and Asanuma, 1972; Asanuma, 1975; and see Welt et al., 1967).

Anatomical Demonstration of Functional Columns

The demonstration of functional columns by electrophysiological methods stimulated the search for morphological correlates of functional columns. The first demonstration of functional columns using anatomical methods was provided by Hubel and Wiesel (1969, 1972), who used the Fink-Heimer technique to show ocular dominance columns in the visual cortex of the monkey. Subsequently, ocular dominance columns were demonstrated by the transneuronal transport of radioactive substances injected into the eye (Shatz et al. 1977; LeVay et al., 1978) and by a reduced silver stain for normal fibers (LeVay et al., 1975). Orientation columns have been observed in the visual cortex of the monkey (Hubel et al., 1978), tree shrew (Skeen et al., 1978), and cat (Albus, 1979; Löwel et al., 1987), by autoradiography of cortical tissue following systemic injections of radioactively tagged 2-deoxyglucose (2-DG). 2-DG is a glucose analogue that labels metabolically active cells, which, in these instances, were activated by specifically oriented stimuli.

Patterns of multicellular arrays in the monkey visual cortex, having similar ocular dominance and orientation selectivity, have been revealed with a noninvasive method employing voltage-sensitive fluorescent dyes (Blasdel and Salama, 1986). A pattern of bloblike structures, extending through layers II and III and appearing more faintly in layers IVB, V, and VI, has been observed in primary visual cortex stained for the mitochondrial enzyme cytochrome oxidase (e.g., Horton and Hubel, 1981; Livingstone and Hubel, 1984a, b; Wong-Riley, 1979). Cells situated between blobs (interblob neurons) respond best to specifically oriented stimuli, and only a minority of these cells are selective for the color of the stimuli, whereas cells within blobs are usually color coded, and only a minority are orientation selective (Livingstone and Hubel, 1984b).

Anatomical correlates of functional columns have been observed also in the somatosensory cortex. For example, in layer IV of the primary somatosensory area of certain rodents, neurons are arranged in three-dimensional multicellular units, called barrels, which may be visualized in both stained and unstained thick sections (Woolsey and Van der Loos, 1970). Electrophysiological recordings have shown that neurons within a single barrel and directly above and below the barrel respond preferentially to stimulation of a single vibrissa on the animal's snout (e.g., Simons, 1978; Welker, 1976). Confirmation that barrels are the morphological manifestation in layer IV of functional columns stretching throughout the full thickness of the cortex has been provided using the 2-DG method (e.g., Durham and Woolsey, 1977; Hand, 1981). Columns of cells having similar receptive field properties have also been labeled in the monkey somatosensory cortex, using the 2-DG method (Juliano and Whitsel, 1987).

Receptive Field Properties of Cortical Neurons

Studies of the receptive field properties of cortical neurons have focused mainly on neurons in the visual cortex. This bias toward visual cortex is most likely due to the ease with which visual stimuli can be manipulated and reproduced and the relative precision with which responses of cortical neurons to these stimuli can be analyzed. The emphasis placed on studying the receptive field properties of visual neurons may also stem from the importance of visual information to vertebrates in general and to humans in particular. In recent years the somatosensory system has received increased attention with the development of methods by which accurate and reproducible stimulation of specific sensory submodalities can be achieved (e.g., Favorov et al., 1987; Gardner and Costanzo, 1980a; Simons, 1983). Because the methods of stimulation, recording, and data analysis employed in electrophysiological studies of visual and somatosensory cortex are somewhat different, results from studies of the visual and somatosensory cortical areas will be considered separately.

Receptive Field Properties of Somatosensory Neurons

In the somatosensory system, the receptor sheets consist of specialized receptors distributed in the skin, the muscle spindles, and the joint capsules, and in specialized dermal structures, like hair follicles (reviewed by Mountcastle, 1979). The receptive field of a neuron in the sensory areas of the cerebral cortex is defined as that area of the receptor sheet where stimulation influences the firing of the cell. The success with which electrophysiological methods can be used to decipher the circuits in which cortical neurons are involved depends on identifying characteristics of the receptive field properties of these cells. One of these characteristics is the size of the receptive field. The receptive fields of different neurons in the cortex have different sizes depending on the area of the periphery represented (e.g., Iwamura and Tanaka, 1978; Mountcastle, 1957); that is, there is an inverse linear relationship between the size of the receptive field and the area of cortex representing a body surface (Sur et al., 1980).

An additional parameter for functional classification of neurons in the somatosensory cortex is the response profile of a neuron to specific stimuli. Some neurons activated by tactile stimulation respond primarily to a change in the stimulus (rapidly adapting, or dynamic neurons), whereas others respond continuously to a steady stimulus (slowly adapting, or static neurons) (Dykes and Gabor, 1981; Mountcastle, 1957; Sur et al., 1984). Jennings (1979) combined extracellular recordings with intracellular injections of HRP and concluded that most rapidly adapting neurons are pyramidal cells, whereas slowly adapting neurons are nonpyramidal. The rapidly adapting neurons can be further divided into two subclasses according to their sensitivity to either high- or low-frequency vibrations (Bennett et al., 1980). Sur et al. (1980) classified somatosensory neurons as having either center-surround receptive fields with separate ON (excitatory response to a stimulus) and OFF (excitatory response to stimulus removal) subregions or uniform receptive fields with no spatially segregated subregions.

A small proportion of the neurons in the somatosensory cortex is unresponsive to simple tactile stimuli; these are considered to have "complex" receptive fields (e.g., Gardner and Costanzo, 1980b; Whitsel et al., 1972). Neurons with complex receptive fields are usually directionally specific; that is, they respond preferentially to a stimulus moving in a specific direction, and some of them respond only to a stimulus presented at an optimal angle (e.g., Constanzo and Gardner, 1980).

Notwithstanding findings showing that neurons in the somatosensory cortex respond to the activation of submodality-specific receptors (e.g., Mountcastle, 1957; and see review by Dykes, 1983), studies using electrical stimulation techniques have revealed a convergence of inputs from several submodalities onto individual somatosensory neurons (Kang et al., 1985; Zarzecki and Wiggin, 1982; Zarzecki et al., 1983). Findings that overlapping regions of the skin may be represented by neighboring or even spatially separate columns (Favorov and Whitsel, 1988) further confound attempts to elucidate cortical circuitry.

In the barrel cortex of mice and rats, many neurons respond to deflections of only one mystacial vibrissa. These neurons respond preferentially either to transient vibrissal movements or to sustained vibrissal displacement at particular angles (e.g., Ito, 1985; Simons 1978; Welker, 1976). The receptive fields of barrel neurons that respond to stimuli from more than one vibrissa consist of separate inhibitory and excitatory subregions, and the receptive field properties of these cells are shaped by the spatial and temporal integration of inputs from the different vibrissae (Simons, 1985). Ito (1985) observed that certain barrel neurons are more sensitive to the amplitude of whisker displacement, whereas other neurons are specialized for detecting the velocity of the displacement. Simons (1978) differentiated between barrel neurons responding to vibrissal stimuli with either fast spikes (having a rapid time course) or slow spikes (having a slow time course).

Available evidence suggests that fast-spiking units are nonpyramidal neurons and that these cells utilize the inhibitory transmitter GABA (cf. Simons et al., 1989). Fast-spiking units in the barrel cortex, unlike most regular-spiking units, can be entrained by relatively high-frequency vibration of the vibrissae; that is, the firing patterns of these units are phase-locked to the vibratory stimulus (Simons, 1978). Thus, these cells, like their counterparts in the primary somatosensory cortex of the monkey (Mountcastle et al., 1969) and in the secondary somatosensory cortex of the cat (Ferrington and Rowe, 1980), may be involved in coding frequency discrimination at a relatively wide range of stimulus velocities.

The diversity of stimulus and response parameters employed by different investigators to describe the receptive field properties of neurons in the somatosensory cortex has impeded attempts to classify somatosensory neurons into distinct functional groups. In addition, the use of anesthesthetic compounds has been shown to affect the receptive field properties of neurons in the somatosensory cortex. For example, as compared with anesthetized rats, neurons recorded from the somatosensory cortex of awake rats have higher spontaneous firing rates, have larger and more complex receptive fields, respond more intensively to stimuli, and respond to activation of submodality-specific receptors that, under anesthesia, do not activate the neurons (e.g., Armstrong-James and George, 1988; Chapin and Lin, 1984).

A potential source of difficulty for classifying neurons according to their response properties is the fact that a large proportion of cortical neurons are unresponsive to stimuli under normal experimental conditions, and thus, these cells are usually excluded from analysis. For example, Dykes and Lamour (1988) have shown that more than half of the cells in the somatosensory cortex of cats and rats have receptive fields that can be detected only following iontophoretic application of depolarizing substances.

These factors have complicated attempts to reveal the synaptic circuitry involved in the formation of the receptive field properties of neurons in the somatosensory areas of the cortex. The stimulus and response parameters employed for the classification of the receptive field properties of neurons in the visual cortex have been more consistent, but as we shall see, findings from investigations using different stimulus and response parameters also have resulted in discrepancies in the classification of the receptive field properties of visual cortical neurons (see Henry, 1977; Stone et al., 1979).

Receptive Field Properties of Visual Neurons

The receptive field of a cell in the visual system is defined as the region of the retina (or visual field) that when stimulated influences the firing of the cell. The majority of neurons in the visual cortex display particular receptive field properties in response to specific visual stimuli, a phenomenon used by Hubel and Wiesel (e.g., 1977) to define classes of cells in the visual cortex of the cat and monkey according to their receptive field properties. *Circularly symmetric* cells, which were identified only in the monkey, show no preference to the orientation of the stimulus and share several receptive field properties with cells in the lateral geniculate nucleus of the thalamus, which provides the major extrinsic input to the visual cortex (Hubel and Wiesel, 1961, 1968). 'Simple' cells have receptive fields with distinct excitatory and inhibitory subregions, each showing summation within the subregion and antagonism to the other subregion. The receptive field map of a simple cell usually consists of elongated ON and OFF regions arranged in parallel, or an elongated region of one type flanked on both, or on one side by the other type of region (e.g., Hubel and Wiesel, 1962, p. 111). In addition, most simple cells respond preferentially to the specific shape, position, and orientation of stimuli producing maximum response. *Complex* cells have relatively large receptive fields with no spatially disparate subregions (e.g., Hubel and Wiesel, 1965, 1968). Thus, complex cells respond to moving stimuli with sustained firing over most of the receptive field, whereas simple cells respond only over a very narrow boundary separating the excitatory and inhibitory regions.

Complex cells were subsequently divided into two subclasses: *standard complex* cells, which respond more strongly as the length of the stimulating slit is increased to the full length of the receptive field, and *special complex* cells, which respond equally to slits of various lengths (e.g., Gilbert, 1977; Palmer and Rosenquist, 1974). An additional class of cells, termed *hypercomplex,* display *end-inhibition,* in which the frequency of response is reduced as the stimulus is elongated beyond an optimal length. Other receptive field properties of hypercomplex cells may resemble those of either simple or complex neurons. It has been suggested (e.g., Dobbins et al., 1987) that hypercomplex cells are specifically involved in the analysis of the curvature of visual stimuli.

Similar functional classifications have been used to define cortical neurons in the visual cortices of other mammals, including the mouse (Dräger, 1975; Mangini and Pearlman, 1980), rat (Burne et al., 1984; Parnavelas et al., 1981, 1983), rabbit (Chow et al., 1971), tree shrew

(Kaufman et al., 1973), mink (LeVay et al., 1987), and opossum (Rocha-Miranda et al., 1973). That this has been possible is no doubt related to the similarity in the organization of the cerebral cortex in different species (see Part I). However, comparisons between the functional properties of visual cortical neurons in different species have revealed dissimilarities in the proportions of cells belonging to the various functional groups, as well as in their laminar distribution.

For example, in the cat visual cortex the majority of neurons have simple receptive fields, and these are found mainly in the deep portion of layer III and in layers IV and VI (e.g., Lennie, 1980). However, in the rat visual cortex, the majority of cells have complex receptive fields, and their laminar distribution is more homogeneous (Parnavelas et al., 1981, 1983). In addition, a significant proportion of the neurons in the visual cortex of the rat and mouse demonstrate no orientation preferences (Burne et al., 1984; Dräger, 1975), whereas nonoriented neurons are relatively infrequent in the visual cortex of the cat (e.g., Hubel and Wiesel, 1962). In the visual cortex of the mink, most cells are driven by inputs from the contralateral eye, whereas in the cat visual cortex, the number of cells dominated by each eye are almost equal; there are also differences in the laminar distribution of the different ocular-dominance cells in these species (LeVay et al. 1987). LeVay et al. (1987) have revealed columns of neurons with similar contrast preferences in the visual cortex of the mink, but not of the cat. They suggested that the differences between these species reflect differences in the functional organization of the lateral geniculate nuclei of these species. Similarly, findings that the proportion of complex cells in rats and mice is significantly higher than in cats may be attributed to differences in the functional organization of the visual system in these animals. It may be that, in rodents, cells with "simple" receptive field properties occur earlier in the visual pathway (e.g., in the LGN or in the superior colliculus), and thus their visual cortices contain a higher proportion of complex cells than in the cat. However, the degree to which these differences may reflect differences in experimental protocols or the difficulty in recording from the relatively small cells in the rodent cortex is unclear.

In any event, these quantitative interspecies differences have not been correlated with any qualitative differences in the receptive field properties of cortical neurons in different species. In fact, that cortical neurons in the visual cortices of all species examined can be classified into similar functional groups supports the proposition, developed in Part I, that the cerebral cortex of different species is built according to some common, basic plan.

A major turning point in our understanding of cortical function was the formulation by Hubel and Wiesel (see review, 1977) of the concept of hierarchical processing of visual information. Hubel and Wiesel noted that distinctive receptive field properties of neurons can be identified at all stages of visual information processing. In the retina (Kuffler, 1953) and in the lateral geniculate nucleus (Hubel and Wiesel, 1961), neurons can be distinguished as having either ON or OFF center-surround receptive fields. According to the proposed hierarchical concept, the receptive fields of simple cells, and of the 'circularly symmetric' cells in the monkey, are formed by the convergence of inputs from geniculocortical afferents, which display either ON- or OFF-center receptive fields. The receptive field properties of complex cells are produced by the convergence of inputs they receive from simple cells.

The hierarchical model of cortical processing has been criticized on various grounds. For example, Henry (1977) argued that both the nomenclature and the functional criteria employed to classify receptive field properties were formulated with a preconception of a hierarchical processing mechanism. Thus, it is not clear whether the response characteristics exhibited by each of the functional cell groups provides evidence for the validity of the hierarchical system or whether the functional cell classes have been formulated so as to conform to this interpretation of cortical processing. For this reason the adoption of this classification method allows for little in the way of alternative interpretations of the synaptic organization in the cerebral cortex.

To rectify this deficiency, Henry proposed an alternative system for classifying neurons in the visual cortex (Henry, 1977; and see Table 1 in Henry et al., 1979). His aim was to develop a classification system that would operate in the absence of any prior commitment to a particular scheme of cortical organization. According to Henry's classification, the "simple" cell is replaced by the S cell, which is distinguished by virtue of its characteristic response properties to both moving and stationary stimuli: (1) Its receptive field consists of one or more discrete areas of ON or OFF discharge; (2) responses from ON and OFF regions, and the responses to moving light and dark edges peak at single points; and (3) lengthening the optimally oriented stimulus increases the response as long as the stimulus remains within the discharge region; lengthening the stimulus beyond the discharge region may diminish, but not abolish the response. The 'complex' cell is replaced by the C cell, which can be differentiated from the S cell by its characteristically uniform receptive field, in which mixed ON and OFF discharges are produced at all points in response to moving and stationary stimuli. The subclasses of complex cells, that is, the standard and special complex cells are denoted with the additions of the subscripts ST and SP, respectively, to the letter C.

Henry's classification includes two additional classes of neurons, the A and B cells. The A cell is similar to the S cell in that it has parcellated ON and OFF regions in the receptive field, yet in the A cell's receptive field, maximum discharge may be evoked at many points across the discharge zone. The B cell is similar to the C cell in several response properties but they differ, for example, in that B cells have smaller receptive fields and exhibit a sustained response to visual stimuli; C cells usually show transient responses. The suffix H added either to the S, C, B, or A denotes that the property of end-zone inhibition ("hypercomplex," according to Hubel and Wiesel's classification) is displayed by a cell belonging to one of the above classes. With this taxonomy, it is possible to assign names for additional functional classes and subclasses of neurons, independent of the existing ones.

A thorough discussion of the limitations and merits of each of the-classification systems described above is beyond the scope of this book, and for this the reader is referred to reviews by Henry (1977, 1985). However, since a major portion of this chapter concerns the correlation of functional classes of neurons with the anatomically defined cell types described in Chapter 2, it would seem useful to point out some of the discrepancies between the different classification methods employed to functionally categorize cortical neurons.

Essentially, the criteria introduced by Hubel and Wiesel for the functional classification of visual cortical neurons were retained in subsequent classification schemes. In every classification scheme, simple cells have spatially discrete zones of ON and OFF discharge, whereas complex cells have uniform receptive fields with intermingled ON/OFF zones. Thus, for example, there is likely to be a close correlation between simple and S cells and between the C and complex cells with uniform receptive fields. However, included in more recent classification schemes are additional stimulus parameters and response properties (c.f. Kato et al., 1978; Pettigrew et al., 1968; Schiller et al., 1976a, b, c; Singer et al., 1975).

This diversity of classification protocols has resulted in numerous discrepancies in the literature, which can be highlighted by considering the classification of hypercomplex cells according to various schemes. In different studies, hypercomplex cells were either not recognized (Hubel and Wiesel, 1962), recognized but excluded from analysis (Burr et al., 1981; Sherman et al., 1976), considered as a subclass of simple or of complex cells (Gilbert, 1977; Martin and Whitteridge, 1984a; Singer et al., 1975), or considered as a distinct class of neurons (Kato et al., 1978; Kelly and Van Essen, 1974).

Because a large proportion of neurons in the visual cortex display the property of end inhibition (e.g., Gilbert, 1977; Kato et al., 1978), it can be assumed that the proportions of simple, complex, and hypercomplex cells reported in the different studies were affected significantly by these different approaches. In addition, it is worthwhile noting that the classification of receptive field properties of cortical neurons is dependent in large part on the stimulation parameters utilized (e.g., Spitzer and Hochstein, 1985). This implies that using different stimulus parameters in different studies may have produced different response characteristics and, ultimately, different classifications for similar neurons. Under these circumstances, it would be difficult to make direct comparisons between the different classification systems used to elucidate mechanisms of cortical information processing.

An additional factor that may complicate the classification of neurons in visual cortex is that in different studies neurons were recorded from experimental animals under different states of anesthesia. As in the somatosensory cortex (see the "Receptive Field Properties of Somatosensory Neurons" section), anesthesia has been shown to affect the receptive field properties of neurons in the visual cortex. For example, anesthesia has been shown to decrease the spontanous activity of cells in the visual cortex, to alter their receptive field size and organization, and to affect their direction and orientation preferences (e.g., Creutzfeldt et al., 1987; and see Bauer and Fischer, 1987). In addition, some neurons in the primary visual cortex (Haenny and Schiller, 1988) and most neurons in the extrastriate cortex (area V4; Haenny and Schiller, 1988; Mountcastle et al., 1987; Spitzer et al., 1988) respond to visual stimuli that require a detection or discrimination task with larger and more selective responses as compared with their responses to no-task stimuli. Modulations of receptive field properties by attentive behavior have also been demonstrated in the auditory (Miller et al., 1972) and in the somatosensory (Hyvärinen et al., 1980) areas of the cortex.

These findings indicate that the response properties of neurons in the cerebral cortex are dependent on the attentive state of the animal, a factor that was not considered in earlier studies of the receptive field properties of neurons in the visual cortex.

According to Hubel and Wiesel's classification scheme, the receptive field properties that characterize simple cells are defined in response to stationary visual stimuli (e.g., Hubel and Wiesel, 1962, 1968). In contrast, other researchers have defined simple cells in terms of their response properties to moving stimuli (Bishop et al., 1971a, b, 1972, 1973; Pettigrew et al., 1968). When presented with an optimally oriented, moving light-dark border, or slit, cortical visual neurons respond only to the edges of the slit, each slit having a leading (light) and a trailing (dark) edge (Bishop et al., 1971a). Cells defined according to Hubel and Wiesel's criteria as having simple receptive field properties respond either to the leading or to the trailing edge, or to both edges of a narrow slit moving through the receptive field, and each cell responds to movement of the slit in either one or both directions (Bishop et al., 1971a; Schiller et al., 1976a).

It appears, then, that several different types of receptive field properties can be anticipated for cells that have all been identified as simple cells according to the classification scheme of Hubel and Wiesel. In addition, cells responding to both edges of the moving slit display two or more firing patterns as the stimulus passes through their receptive fields. The activity profiles of these cells are dependent on various stimulus parameters, such as the slit width and the direction of movement, factors that introduce additional complications in the organization of the receptive fields (Bishop et al., 1971a). Thus, the systematic use of moving stimuli has revealed a complexity of receptive field organization that is not apparent when receptive properties of neurons are defined according to their responses to stationary stimuli.

Using relatively complex stimulus parameters to study receptive field properties has revealed additional receptive field features of neurons in the visual cortex. For example, Gulyás et al. (1987) showed that neuronal responses to a moving bar are suppressed with the simultaneous presentation of a textured background that moves at the same speed or faster than the bar. Furthermore, use of the moving textured background paradigm, which may be considered to more closely resemble natural visual stimuli, has revealed six different types of directionally specific cortical neurons, each type responding preferentially to the relative direction and velocity of the moving bar against the moving background (Orban et al., 1987).

Additional receptive field properties of neurons have been revealed by using colored stimuli; some neurons respond selectively or preferentially to stimuli having particular wavelengths, and some of these cells have receptive fields consisting of mutually antagonist subregions, each responding to stimuli of different wavelengths (e.g., Michael, 1978, 1987; Livingstone and Hubel, 1984b, Hubel and Livingstone, 1987). Some neurons in the visual cortex are characterized by an additional receptive field property, termed *disparity tuning,* in which their responses are critically dependent on the relative horizontal positions of the stimuli in the two eyes (reviewed by Hubel and Livingstone, 1987).

These findings imply that classifying neurons according to their responses to single, stationary, monochromatic stimuli provides an incomplete picture of the receptive field properties of neurons in the visual cortex (see above [classification of the receptive field properties of neurons in the somatosensory cortex using restricted stimulus parameters]).

The wide spectrum of receptive field properties revealed by employing various stimulus parameters implies that any single system of cell classification may be arbitrary. Clearly, the classification of neurons based on the hierarchical model does not account for many of the receptive properties displayed by cortical neurons, a finding that implies that the hierarchical model by itself cannot provide a thorough description of the mechanisms of cortical information processing.

This theme will be taken up again in Part III, in which hierarchical and parallel models are compared. The remainder of Part II will be devoted to reviewing data on various mechanisms that may be responsible for the formation of receptive properties of cortical neurons. The emphasis on receptive fields and on the mechanisms that shape them stems from the presumption that the receptive fields of neurons reflect their synaptic connectivity. Thus, studies of the mechanisms shaping the receptive field properties of cortical neurons can be expected to provide valuable information on the synaptic connectivity of these cells.

Laminar Distribution of Functionally Defined Neurons

Thalamic afferents terminate preferentially in layer IV of the primary sensory areas (Chapters 1 and 3). A large proportion of the neurons in layer IV are nonpyramidal (e.g., Peters et al., 1985), and most of the units in this layer have simple receptive fields properties (Gilbert, 1977; Hubel and Wiesel, 1962, 1968). In the supra- and infragranular layers, pyramidal cells predominate (e.g., Peters et al., 1985), as do cells having more complex receptive field properties (Gilbert, 1977; Hubel and Wiesel, 1962, 1968). For this reason, it was proposed that simple cells in layer IV are nonpyramidal neurons, whereas complex cells in the superficial and deep layers of the cortex are pyramidal neurons (e.g., Gilbert, 1977; Hubel and Wiesel, 1962, 1968).

Findings consistent with these speculations have been provided by additional studies that have correlated the receptive field properties of neurons with their laminar distribution. For example, Henry et al. (1979) report that cells with nonoriented receptive fields are found in layer IV of the cat visual cortex; the majority of S cells are concentrated in layers IV and VI, and B and C cells predominate in layers III and V.

Regardless of the classification method employed, it is generally agreed that most of the cells in layer IV have simple receptive properties, whereas only a small proportion of the cells in this layer have complex receptive field properties (e.g., Berman, 1982; Kato et al., 1978; Martin and Whitteridge, 1984a; Schiller et al., 1976a). In addition, Hubel and Wiesel (1965) noted that simple cells are found most often in the primary visual cortex (Brodmann's area 17), the major site for geniculocortical input, whereas areas 18 and 19, which receive less thalamic input, contain a large proportion of complex and hypercomplex cells, the latter being preferentially distributed in area 19.

The receptive field properties of neurons have been correlated with their laminar distribution in the somatosensory cortex also. For example, in the primary somatosensory cortex of the rat (Chapin, 1986; Simons, 1978), mouse (Simons and Woolsey, 1979), and monkey (Sur et al.,

1985), neurons with the smallest receptive fields are found in layer IV, whereas neurons above and below layer IV have progressively larger receptive fields. In addition, slowly adapting neurons are found only in the middle cortical layers of the monkey somatosensory cortex, but rapidly adapting neurons are found in all cortical layers (Sur et al., 1984). In the rat barrel cortex, fast-spiking neurons are largely restricted to layer IV, whereas regular-spiking neurons are found throughout layers II to VI (Simons, 1978). Also, most neurons in layer IV of the rat barrel cortex respond to deflections of only one vibrissa, whereas neurons in the supragranular and infragranular layers, respectively, are activated by deflections of progressively larger numbers of vibrissae (Armstrong-James and Fox, 1987; Simons, 1978).

Nevertheless, there are significant limitations in correlating receptive field properties with laminar location, because every cortical layer contains a mix of morphological types. Unless the recorded neurons are labeled, meaningful correlations between receptive field properties and morphology are not possible.

Functional Properties of Morphologically Identified Neurons

An effective approach to correlating the function and structure of cortical neurons involves injecting neurons with fluorescent dyes (e.g., Kelly and Van Essen, 1974; Van Essen and Kelly, 1973) or with HRP (e.g., Martin and Whitteridge, 1984a) after their receptive field properties have been defined. Intracellular injection of HRP has the advantage that the fine structure and the synaptic connections of the HRP-filled neurons can be examined with the electron microscope (e.g., Martin et al., 1983). Studies dealing with receptive field properties of identified cortical neurons are few, and these focus almost exclusively on the visual cortex. Parnavelas (1984) estimated that during the last two decades structure/function correlations have been made for fewer than 200 neurons, and in only half of these instances have their morphological and functional properties been described in detail. The vast majority of the cells studied with these intracellular labeling methods are pyramidal neurons, a bias most likely due to the preponderance of this cell type in the visual cortex and to the relatively large size of their cell bodies compared with most nonpyramidal neuronal types (e.g., Peters et al., 1985).

Figures 4.1 to 4.3 provide quantitative data from the cat visual cortex on 181 neurons whose receptive field properties were correlated with their morphology and laminar distribution. Cells in the simple category include units identified as simple cells according to Hubel and Wiesel's classification, as well as units identified as S cells according to Henry's classification. Similarly, the complex category includes both Hubel and

Wiesel's complex cells and Henry's C and B cells. Figure 4.1 describes the laminar distribution of morphologically and functionally defined cells, and Figures 4.2 and 4.3 describe the correlation between the morphologies and the receptive field properties of these cells.

Receptive Field Properties of Identified Neurons

Pyramidal Neurons. Pyramidal neurons in the visual cortex of the cat occur in about equal proportions as complex or simple receptive field types (Figure 4.2). Pyramidal cells with complex receptive fields are distributed mainly in layers II, III, and V; the receptive fields of pyramidal cells in layer V are typically larger than those of pyramidal cells

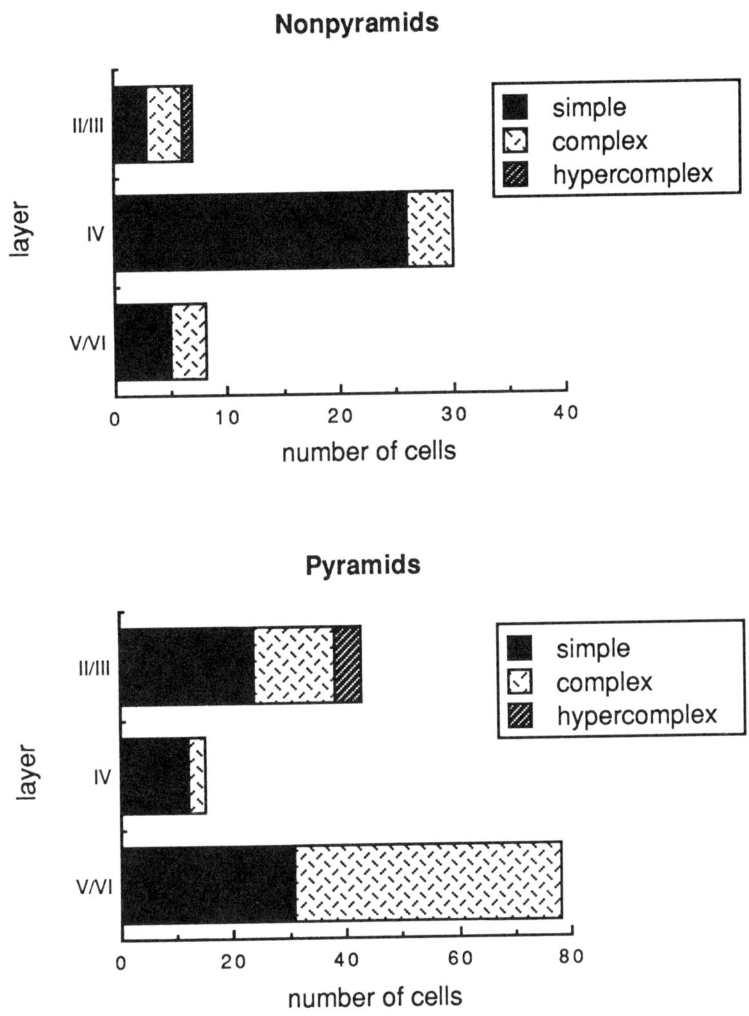

Figure 4.1. *The morphologies, laminar distribution, and receptive field properties of neurons in the cat visual cortex. Cumulative histograms showing the proportions of cells with simple and complex receptive fields within different layers of the cat visual cortex. The data are derived from several studies (Kelly and Van Essen, 1974; Gilbert and Wiesel, 1979, 1981; Lin et al., 1979; Martin and Whitteridge, 1984a), in which a total of 181 neurons were intracellularly filled with HRP or a fluorescent dye after their receptive field properties had been determined. A small number of neurons (<2%) from these studies were not included because insufficient details were provided on their morphological and/or receptive field characteristics. The number of hypercomplex cells is grossly underestimated, because many cells that showed end inhibition were classified in these studies as simple or complex neurons.*

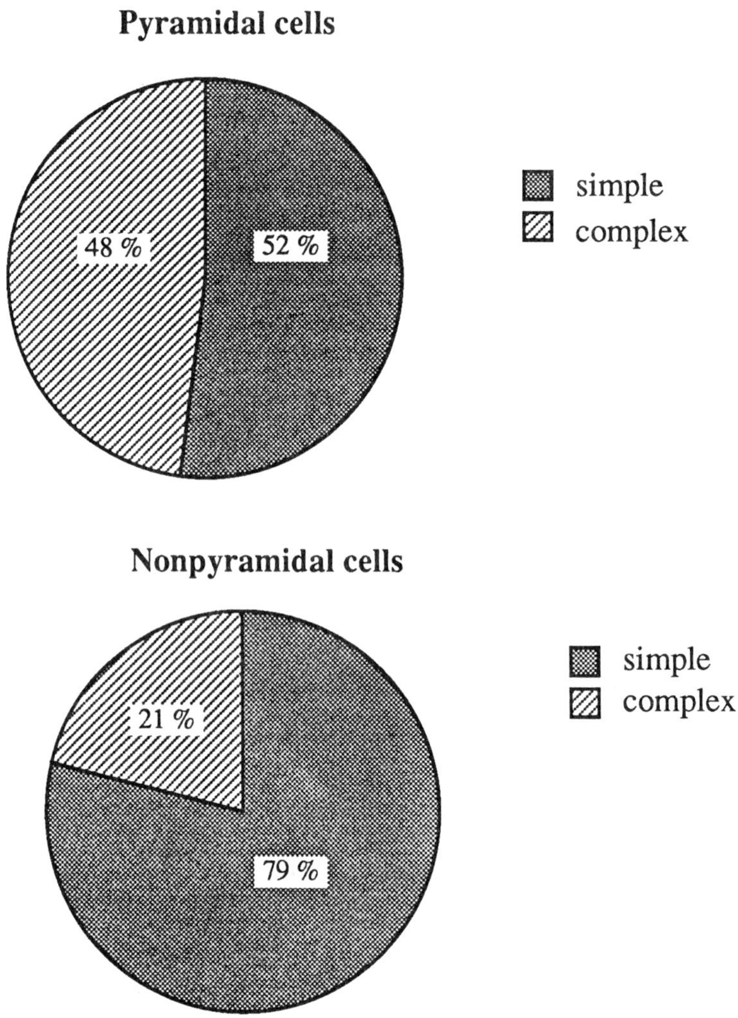

Figure 4.2. *The proportion of pyramidal and nonpyramidal neurons in the cat visual cortex having simple or complex receptive fields.* Pyramidal cells: $n = 131$; nonpyramidal cells: $n = 44$. *Hypercomplex cells were omitted for the sake of clarity, since they comprise less than 4% of the total cells examined. See legend of Figure 4.1 for additional details.*

in the superficial layers (Gilbert and Wiesel, 1979, 1981; Kelly and Van Essen, 1974). Several studies have reported that the majority of pyramidal neurons in layers II and III have complex receptive field properties (e.g., Gilbert and Wiesel, 1979; Kelly and Van Essen, 1974). A different finding was reported by Martin and Whitteridge (1984a), who found that the majority of pyramidal neurons in layers II and III have S type receptive fields. One possible explanation for this contradiction is that Martin and Whitteridge (1984a) considered end inhibition as a property of either simple or complex cells, whereas Kelly and Van Essen (1974) classified cells having this property as hypercomplex cells. Alternatively, differences in the proportions of simple and complex py-

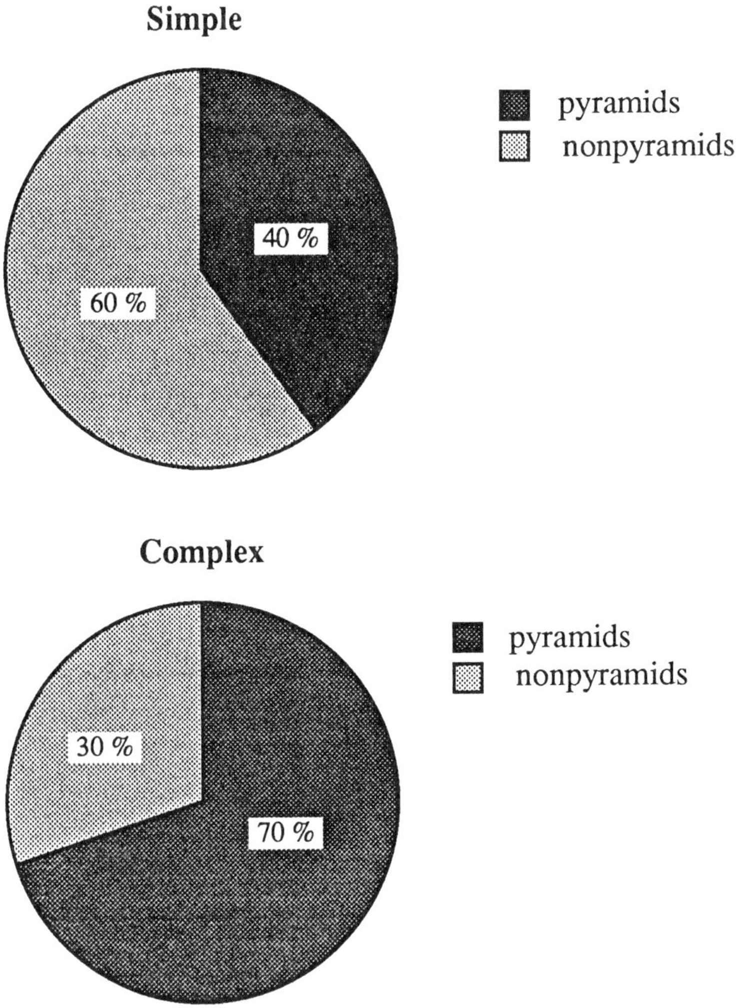

Figure 4.3. *The proportion of simple and complex cells in the cat visual cortex having pyramidal or nonpyramidal morphologies. Data corrected for sample size.* Simple cells: n = 101; complex cells: n = 74. *Hypercomplex cells were omitted for the sake of clarity, since they comprise less than 4% of the total cells examined. See legend of Figure 4.1 for additional details.*

ramidal neurons reported in these studies may stem from the fact that Martin and Whitteridge (1984a) recorded from 26 pyramidal neurons in layers II/III, whereas Kelly and Van Essen's (1974) data are derived from a sample of 8.

Approximately 60% of the pyramidal neurons in layers V and VI of the cat visual cortex have complex receptive field properties; the remainder have simple receptive fields (Gilbert and Wiesel, 1979, 1981; Kelly and Van Essen, 1974; Lin et al., 1979; Martin and Whitteridge, 1984a; and see Figure 4.1). The majority of pyramidal neurons in layer IV have simple receptive field properties (e.g., Martin and Whitebridge, 1984a; and see Fig. 4.1). In the rat visual cortex, pyramidal neurons with complex receptive field properties have been identified in layers II through VI, whereas a few pyramidal cells with either simple or hypercomplex receptive fields have been identified in layers II, III, and V (Parnavelas et al., 1983).

Nearly all pyramidal cells send their axons to other areas of the brain (see Chapter 2), and attempts have been made to correlate the projection sites of these cells with their receptive field properties. In Chapter 3 the proposition was advanced that corticofugal cells projecting to specific locations receive characteristic sets of afferent inputs. If that is so, it is conceivable that different types of projection neurons would have different and characteristic receptive field properties. It has been reported that most corticotectal (Harvey, 1980; Singer et al., 1975; Swadlow and Weyand, 1987; Weyand et al., 1986), corticocollicular (Palmer and Rosenquist, 1974), and corticopontine cells (Albus and Donate-Olive, 1977) display complex receptive field properties. However, other classes of corticofugal cells may contain neurons having either simple or complex receptive field properties. Among these are corticostriatal (Henry, 1977), corticothalamic (Gilbert, 1977; Harvey, 1980; Singer et al., 1975), and transcallossal neurons (Harvey, 1980; McCourt et al., 1987; Simmons and Pearlman, 1983). These findings imply that corticofugal cells projecting to a single target area do not necessarily receive similar inputs.

Nonpyramidal Neurons. The majority of nonpyramidal neurons in the cat visual cortex have simple receptive field properties (Figure 4.2). Nonpyramidal neurons with simple receptive fields occur mainly in layer IV and include spiny stellate cells (Gilbert and Wiesel, 1979; Lin et al., 1979; Martin and Whitteridge, 1984a), nonspiny, and sparsely spiny nonpyramidal neurons with spherical or elongated dendritic fields (Gilbert and Wiesel, 1979; Kelly and Van Essen, 1974). Martin and Whitteridge (1984a) report that all spiny stellate cells, which occur exclusively in layer IV, have S-type receptive fields. A small number of nonpyramidal neurons in layers II, III, V, and VI have either simple, complex, or hypercomplex receptive field properties (Gilbert and Wiesel, 1979; Kelly and Van Essen, 1974; Martin and Whitteridge, 1984a; and see Figure 4.1). Nonpyramidal neurons of layers III to VI, with either S- or C-type receptive fields (Martin and Whitteridge, 1984a), have been identified as basket or clutch cells (Somogyi et al., 1983a), and some were shown by immunocytochemical methods to contain GABA (Kisvarday et al., 1985, 1987; Somogyi and Soltész, 1986). In the visual cortex of the rat, three functionally defined nonpyramidal neurons have been studied. Two of these cells were situated in layer IV and possessed simple receptive fields, whereas the third cell, a layer VI sparsely spinous multipolar neuron, featured "ON–OFF" receptive field properties (Parnavelas et al., 1983).

The data presented above and illustrated in Figures 4.1 to 4.3 indicate that only a general correlation exists between receptive field properties and neuronal morphology: The majority of complex cells are pyramidal neurons, whereas the majority of simple cells are nonpyramidal neurons (Figure 4.3). Viewed the other way around, most nonpyramidal neurons have simple receptive field properties, but pyramidal neurons are almost evenly split between those with simple and those with complex receptive field properties (Figure 4.2). In addition, cells situated in a single layer of cortex, even when they display similar morphologies or utilize the same neurotransmitter, may exhibit different receptive field properties.

According to the hierarchical model, input from thalamic afferents is responsible for the formation of the receptive field properties of simple cells. The progressively more complex receptive fields of cells at higher levels of the proposed hierarchy are assumed to be formed by inputs from simple cells in layer IV to the more superficial layers, which, in turn provide input to the deeper layers. However, because cells having somata in any layer of the cortex are postsynaptic to thalamocortical afferents (Table 3.2), and in line with the proposition that thalamocortical input is a principal determinant of simple receptive fields, it may be assumed that simple cells occur in every layer of the cortex; the data in Figure 4.1 are consistent with this assumption. Nevertheless, it cannot be excluded that hierarchical processing of information occurs between neurons with progressively more complex receptive properties, even if these neurons are not segregated into distinct morphological subgroups or in specific laminae.

Shape of the Dendritic Tree. Because the receptive field properties of a cell, its laminar distribution, and its morphology cannot be strictly correlated in the cerebral cortex, alternative approaches have been sought to correlate the form and function of cortical neurons. One approach has been to correlate the receptive field properties of neurons with the spatial organization of their dendrites (e.g., Gilbert and Wiesel, 1979). Because axonal pathways are known to terminate preferentially within specific laminae (see Chapter 1), the spatial relationships between specific cortical laminae and the dendritic arborizations of cortical neurons must affect the types and numbers of inputs the neurons are likely to receive. The striking association between specific dendritic arborization patterns and cortical laminae was described in Chapter 2, where it was noted that the pattern of dendritic arborization can be correlated with the laminar location of the cell body and with the projection sites of the cell's axon (see Summary, Chapter 2).

One approach to correlating the receptive field properties of neurons with the shapes of their dendritic trees is based on injections of HRP into neurons whose receptive field properties have been determined. For example, Gilbert and Wiesel (1979) have reported that, in the cat visual cortex, cells with larger dendritic arborizations have larger receptive fields. In contrast, Martin and Whitteridge (1984b; and see Martin, 1984) have found no correlation between the size of the dendritic tree and the size of the receptive field; nor did they find any correlation between the orientation of the dendritic trees and the orientation of the receptive fields.

Correlations between receptive field properties of neurons and the shapes of their dendritic trees have been investigated on the assumption that the shape of the dendritic tree determines whether a cell is likely to receive particular inputs and that these inputs are responsible for the formation of the cell's receptive field properties. However, it is not clear whether the laminar distribution of a neuron's dendrites is the prime determinant of its ability to receive specific inputs. For instance, White and Rock (1981) examined thalamocortical input onto dendrites belonging to two adjacent nonpyramidal neurons in the mouse barrel cortex and found that the proportion of thalamocortical synapses formed with dendrites of one cell was five times greater than the proportion of these synapses formed with dendrites of the other cell, even though all the dendrites examined passed through neuropil containing similar concentrations of thalamocortical synapses. Similarly, Keller and White (1987) report that different GABAergic neurons in the mouse barrel cortex receive markedly different proportions of thalamocortical synapses, even though their dendritic trees have similar orientations and occur in close proximity to one another. Thus, it appears that the specific inputs neurons receive are not determined solely by the laminar distribution and shape of their dendritic trees, but that different cells are somehow predisposed to receive specific numbers and types of synapses from various sources. Part III will include a discussion of specificity in synaptic connectivity.

Biophysical Properties of Identified Neurons

The classification of cortical neurons into functional groups usually is based on the response properties of neurons to various stimuli. However, the functional properties of neurons may depend, at least in part, on various properties intrinsic to the neurons, such as the biophysical properties of their membranes. A large body of data exists on the biophysical properties of cortical neurons, but in the vast majority of these studies, the morphologies and laminar location of the neurons have not been determined, and so it is difficult to relate the results of this work to cortical circuitry.

Some insight into the relationship between the morphology of neurons and their biophysical properties is provided by the finding that pyramidal tract neurons in the motor cortex are composed of neurons having fast and slow axon conduction velocities; these neurons correspond to Betz type cells and to smaller pyramidal neurons, respectively (e.g., Deschênes et al., 1979). Corticopontine pyramidal neurons in the cat parietal cortex include rapidly conducting neurons, which have large somata and aspiny or sparsely spiny apical dendrites, and slowly conducting neurons, which have smaller somata and richly spinous apical dendrites (Yamamoto et al., 1987a). Similar morphological features characterize rapidly versus slowly conducting corticopyramidal and corticopontine pyramidal cells in the somatosensory cortex of the cat (Yamamoto et al., 1987b).

A promising approach to correlating biophysical properties with the morphologies and laminar location of cortical neurons is the brain slice preparation. The technique of recording from brain slices was introduced in the early 1960s (e.g., Hillman et al., 1963) and employed in systematic studies of cortical neurons by Gutnick, Connors, and Prince (e.g., Gutnick et al., 1981; Connors et al., 1981, 1982). Several studies have shown that intracellular recordings can be maintained for several hours from neocortical slices and that the neurons in the slice preparation retain the basic membrane and synaptic properties ascribed to them in situ (e.g., Connors et al., 1982).

A limitation of the neocortical slice preparation is the fact that the physical integrity of the tissue cannot be maintained, and thus, both the afferents and efferents of the recorded cells, as well as parts of the dendritic tree, may be severed when the slice is prepared. One of the major advantages of the brain slice preparation is the relative ease with which neurons can be recorded from intracellularly and injected with neuronal markers. By combining intracellular recording with labeling using either fluorescent dyes or HRP, the morphological and electrophysiological properties of a neuron can be correlated. McCormick et al. (1985) have employed this approach to determine whether pyramidal and nonpyramidal neurons have distinct electrophysiological properties.

McCormick et al. (1985) examined neurons in brain slices of the sensorimotor and anterior cingulate cortices of the guinea pig and identified three distinct cell classes, based on their electrophysiological behavior. The first class, termed *regular-spiking cells,* were characterized by their relatively long action potentials, prominent spike-frequency adaptation, and a response to local stimuli that was characterized by brief excitation followed by prolonged inhibition. All regular-spiking cells were identified as pyramidal neurons, and these occurred in layers II through VI of the cortex. The second class of neurons were termed *bursting cells* because they generated bursts of three to five action potentials. These were also identified as pyramidal neurons having cell bodies in layer IV or in upper layer V. The third class of neurons, termed *fast-spiking cells,* had very short action potentials, large postspike undershoots, little or no adaptation, and a response to local synaptic activation that was characterized by prolonged synaptic excitation. All fast-spiking units were identified as nonspiny or sparsely spiny nonpyramidal neurons and were found in layers II to VI.

Studies of the turtle visual cortex (Connors and Kriegstein, 1986; Kriegstein, 1987), in which physiologically identified neurons were stained with either Lucifer yellow or HRP, revealed a similar dichotomy for the biophysical properties of pyramidal and nonpyramidal neurons. Pyramids were characterized by prolonged action potentials and marked frequency adaptation (regular spiking), whereas nonpyramidal cells exhibited relatively brief action potentials, postspike undershoots, and little or no spike frequency adaptation (fast spiking). Similar classes of neurons, distinguished by their biophysical properties, were revealed in the mouse barrel cortex (Agmon and Connors, 1987). Different findings were reported by Dykes et al. (1988) who recorded from, and labeled, two pyramidal neurons in the somatosensory cortex of the cat; both were fast-spiking cells. It is not yet known whether these conflicting findings are related to interspecies differences or whether more than one morphological class of fast-spiking cortical neurons exist.

These studies suggest that pyramidal and nonpyramidal neurons in the cerebral cortex possess distinctly different biophysical characteristics, which no doubt profoundly affect their responses to peripheral stimuli (see McCormick et al., 1985). For example, regular-spiking pyramidal cells, which show a slow rate of action potential repolarization and a strong adaptation to spike frequency, would be particularly sensitive to transient or phasic excitation. Pyramidal cells that generate bursts of action potentials, but display membrane properties similar to those of regular-spiking cells, should respond vigorously but transiently to abrupt stimuli, but otherwise maintain a sustained output to tonic stimuli. The sparsely spiny and nonspiny cells, with their brief action potentials and little or no adaptation, appear to be well suited to transform a tonic excitation into a sustained output.

Summary of Chapter 4

Neurons in sensory areas of the cerebral cortex respond preferentially to stimuli having specific characteristics. These characteristics, which define the receptive field properties of cortical neurons, include the type and number of peripheral receptors activated by the stimulus; the physical properties of the stimulus, such as shape, size, wavelength, velocity, or frequency; and the mode of stimulus presentation, such as the orientation or direction of movement of the stimulus.

Neurons having different morphological characteristics differ also in their biophysical properties. Studies of the receptive field properties of cortical neurons were critical for the identification of functional columns, which also have been identified using a variety of anatomical techniques. Neurons within a functional column have similar but not necessarily identical receptive field properties. Differences in their receptive field properties may stem from the hierarchical processing of information by different types of neurons in different layers of the cortex.

The hierarchical model was challenged on several grounds, and these will be discussed in detail in Chaper 7. One of the criticisms raised against this model was that the criteria for neuronal classification do not take into account all the receptive field properties of cortical neurons. Combined anatomical and physiological studies have not established a clear correlation between the shapes, laminar distribution, and receptive field properties of cortical neurons.

Elucidation of the mechanisms shaping receptive field properties is contingent upon revealing synaptic interactions between neurons. Cortical circuits, as revealed by electrophysiological methods, are discussed in the following chapter.

5 Synaptic Circuitry Revealed by Electrophysiology

The receptive field properties of cortical neurons are shaped by synaptic inputs from a variety of sources. These sources include extrinsic input from subcortical regions and from both contralateral and ipsilateral areas of the cortex, and intrinsic inputs from the local axon collaterals of both pyramidal and nonpyramidal neurons. The anatomical evidence for these synaptic interactions was discussed in detail in Chapter 3; this chapter presents data obtained with electrophysiological methods that confirm and extend the results obtained by the neuroanatomical studies.

Synaptic interactions between neurons have been examined by stimulating one or more neurons with an electrical current and recording the response properties of other cells at various distances from the stimulating electrode. The action potential produced by the stimulating electrode may travel orthodromically along axons afferent to the neuron at the recording site or antidromically along the axon belonging to the same neuron from which the recordings are taken. Because afferent and efferent projections usually coexist within a small volume of neuropil, difficulties have been encountered in differentiating responses evoked by orthodromic stimulation from those produced by antidromic stimulation. For example, activation of cells in the thalamus induces orthodromic activation of thalamocortical afferents, as well as antidromic activation of axon collaterals belonging to corticothalmic projection cells; responses to both activations are recorded in the same small area of the cortex (e.g., Ferster and Lindström, 1983). To circumvent this problem, methods have been developed by which orthodromic and antidromic stimulation can be produced selectively (e.g., Bishop et al., 1962; Ferster and Lindström, 1983, 1985). Moreover, by determining the time lag between the initiation of the stimulus and the onset of the postsynaptic potential, the number of synapses along a neuronal pathway can be estimated. Thus, an established connection between a pair of neurons may be identified as mono-, di-, or trisynaptic if its time lag falls within a latency range assumed for one, two, or three synaptic connections.

An additional approach by which synaptic interactions between neurons can be identified is the method of cross-correlation analysis, introduced by Perkel et al. (1967). In this method, one electrode is used to record from a reference cell, and a second electrode is placed at varying distances to record from other cells. By recording spontaneous spiking activity from the two cells and by analyzing the time difference between the occurrence of these spikes, it can be determined whether the two cells fire simultaneously or whether the activity in one cell precedes that of the other. By employing electrical or natural stimuli, it can be determined whether two cells receive a common input and whether these cells interact with each other monosynaptically (see Abeles, 1982).

In spite of the technical difficulties involved in recording simultaneously from two neurons and of the difficulty imposed by the fact that most cortical neurons have very low rates of spontanous discharge (e.g., Abeles, 1982), the cross-correlation technique has several advantages. The fact that interactions between cells can be monitored without stimulation avoids the complications introduced by artificial electrical stimuli. In addition, by recording from a selected pair of neurons, synaptic interactions between cells can be revealed that would otherwise be concealed by interactions between large numbers of cells. The cross-correlation approach can determine also whether a synaptic interaction is excitatory or inhibitory and the relative strength of the interaction. Finally, by applying peripheral stimuli, the receptive field properties of the recorded neurons can be identified.

Extrinsic Afferents

A major source of extrinsic input to the cerebral cortex is the thalamus. Since the 1970s, electrophysiological techniques have been combined with neuroanatomical methods in order to correlate the functional properties of thalamocortical projections with the morphologies and the receptive field properties of their postsynaptic targets; nearly all these data are derived from studies of the visual system.

Most neurons in the retina (Kuffler, 1953) and in the lateral geniculate nucleus (LGN) of the thalamus (Hubel and Wiesel, 1961) can be identified as having either ON or OFF center-surround receptive fields. In the visual cortex, certain cells display receptive field properties that are related to the receptive field properties of neurons in the LGN, and these cells are thought to receive direct geniculate input (Hubel and Wiesel, 1962; and see Chapter 4).

Subsequent studies have shown that both ON and OFF center-surround categories can be subdivided into additional functional classes, according to the X/Y/W classification (e.g., Enroth-Cugell and Robson, 1966; Hoffmann et al., 1972; and review by Lennie, 1980). This classification scheme may be summarized as follows: X cells, as compared with Y cells, feature a more sustained response, have smaller receptive field center sizes, are more linear in their spatial summation characteristics, and have axons that exhibit slower conduction velocities. W-type (or sluggish) cells are characterized by their low responsiveness and their slow axon conduction velocity; most W cells project to the superior colliculus, and only a sparse projection of W-type afferents reaches the cortex (see review by Lennie, 1980). Functionally defined X and Y cells have been correlated with distinct, anatomically identified cell classes, both in the retina (Cleland et al., 1975; Fukuda and Stone, 1975) and in the LGN (LeVay and Ferster, 1977; Wilson et al., 1976). In the LGN of the cat, there exist two functional classes of X cells, which are characterized by their responses to visual stimuli with either long latency (lagged X cells) or short latency (nonlagged X cells) discharges (Mastronarde, 1987a). The two classes of X cells, which also differ in other functional features (Humphrey and Weller, 1988a), can be correlated with anatomically distinct cell classes in the LGN (Humphrey and Weller, 1988b).

Retinal ganglion cells with X- or Y-type receptive fields innervate X- or Y-type geniculate cells, respectively, and the geniculocortical afferents retain the specific receptive properties exhibited by their parent cell bodies (e.g., Cleland et al., 1971; Hoffmann et al., 1972; Stone and Hoffmann, 1971). Thus, visual information is relayed from the retina, through the LGN, and to the visual cortex by functionally specific afferents within which the spatial segregation of functional streams is maintained. Findings that different functional classes of X cells exist in the cat LGN, but not in the retina (Mastronarde, 1987b), indicate that visual information relayed by a specific functional stream is processed within the relay stations, before reaching the visual cortex (cf. Humphrey and Weller, 1988a).

It has been postulated that X and Y afferents innervate specific populations of neurons in the cortex (see Henry, 1985). Evidence that this may be so was provided by Freund et al. (1985b), who showed that, in the cat visual cortex, X axons synapse with GABAergic cell bodies that are significantly smaller than those receiving input from Y axons. However, neurons in the visual cortex do not display distinct X- or Y-type receptive fields, and thus, it appears that the receptive field properties of the separate functional streams are transformed by cells in the visual cortex. Understanding the mechanisms underlying this transformation is contingent upon revealing the relationships between thalamic input and the receptive field properties of neurons in the visual cortex.

It has been suggested that X and Y afferents synapse with different neurons in different layers of the cat striate cortex (area 17). Support for this assumption has been provided by anatomical studies, showing that terminals belonging to X- and Y- type geniculate relay cells are segregated within layers IVc and IVab, respectively (Ferster and LeVay, 1978).

Direct evidence for the segregation of X and Y afferents in the cortex has been provided by Gilbert and Wiesel (1979, 1981), who used intracellular recordings and injections of HRP into individual geniculocortical axons. Their findings indicate that X afferents arborize entirely in deep layer IV of the cat striate cortex, whereas Y afferents terminate within mid- to upper-layer IV and occasionally send axon collaterals to layer VI. Other studies indicate that a perfect laminar segregation of X and Y afferent input does not occur. For instance, Humphrey et al. (1985) found in the cat area 17 that the terminal fields of X and Y afferents overlap substantially in lower layer III and throughout layers IV and VI. Similarly, Freund et al. (1985a) observed that both X and Y afferents terminate in upper layer IV and in layer VI.

The overlap in cortical laminae of X and Y afferents does not necessarily mean that these afferents synapse with the same postsynaptic elements, although Rule 1 (Chapter 3) would suggest that this is so. Because each cortical layer contains several neurons having different receptive field properties, the possibility exists that specific neuronal types receive input from only one or the other type of thalamic afferent. This possibility has been examined by correlating the morphologies and receptive field properties of visual cortical neurons with the types of thalamic afferents they receive. Table 5.1 presents data from several studies of the cat striate cortex, in which neurons monosynaptically activated by geniculocortical afferents were identified according to their morphologies and/or their receptive field properties.

That both pyramidal and nonpyramidal neurons in sensory areas of the cortex receive direct thalamic input is consistent with the anatomical data summarized in Table 3.2 (Chapter 3). A similar arrangement has been observed for pyramidal and nonpyramidal neurons in the cat motor (Noda and Yamamoto, 1984) and auditory cortices (Mitani et al., 1985) using combined intracellular recordings and injections of HRP. Indicated by the data in Table 5.1 is the fact that both simple and complex cells in the cat striate cortex receive monosynaptic input from X- or Y-type

Table 5.1. *Neurons in the cat striate cortex that are monosynaptically activated by LGN afferents.*

Cell type	Receptive field type	Type of afferents	Selected references
Unclassified	Simple or complex	?	Toyama et al., 1973, 1977; Henry et al., 1979; Ferster and Lindström, 1983, 1985
	Simple	X and Y	Lee et al., 1977; Tanaka, 1983
	Simple	X	Hoffman and Stone, 1971; Toyama et al., 1973; Singer et al., 1975; Lee et al., 1977; Tanaka, 1983
	Simple	Y	Singer et al., 1975; Lee et al., 1977; Tanaka, 1983
	Complex	X and Y	Tanaka, 1983
	Complex	X	Singer et al., 1975; Lee et al., 1977; Tanaka, 1983
	Complex	Y	Hoffman and Stone, 1971; Toyama et al., 1973; Singer et al., 1975; Tanaka, 1983
Pyramids	Simple	X	Martin and Whitteridge, 1984a
	Simple	Y	Martin and Whitteridge, 1984a
	Complex	X	Martin and Whitteridge, 1984a
	Complex	Y	Martin and Whitteridge, 1984a
Nonpyramids	Simple	X	Martin et al., 1983; Martin and Whitteridge, 1984a
	Simple	Y	Martin et al., 1983; Martin and Whitteridge, 1984a
	Complex	X	Martin et al., 1983; Martin and Whitteridge, 1984a
	Complex	Y	Martin et al., 1983; Martin and Whitteridge, 1984a

afferents. Some simple and complex neurons in the visual cortex may receive convergent inputs from both X and Y afferents (Lee et al., 1977; Tanaka, 1983).

That both simple and complex cells receive direct thalamic input does not necessarily refute the validity of the hierarchical model, which implies that only the receptive field properties of simple cells are formed by direct input from thalamic afferents. It can be assumed that cells receiving a larger proportion of their input from thalamic afferents would have receptive field properties that are more closely related to the receptive field properties of their afferents, whereas cells receiving lesser amounts of thalamic input would have receptive field properties that are less directly related to the receptive field properties of thalamic afferents. Singer et al. (1975) have reported that both simple and complex cells receive direct geniculocortical input: Simple cells are driven almost exclusively by LGN afferents; complex cells are driven, in addition, by intrinsic projections, by recurrent collaterals of corticofugal cells, and by callosal fibers. Thus, differences between the receptive field properties of simple and complex cells in the cat visual cortex may be related to the relative proportion of thalamic input these cell types receive.

Most neurons that project from the thalamus to the barrel cortex of the rat have multiwhisker receptive fields; that is, they respond to inputs from a large number of whiskers (Simons and Carvell, 1989). In comparison, many neurons in the barrel cortex respond preferentially to inputs from only one vibrissa (e.g., Simons, 1978). How, then, are the single-whisker receptive fields of neurons in the barrel cortex generated?

Simons et al. (1989) suggested that cortical neurons receive convergent inputs from a number of thalamic neurons (Figure 5.1). This suggestion is supported by findings showing that single-whisker units in the barrel cortex receive subthreshold, multiwhisker inputs (Carvell and Simons, 1985). When a single whisker is activated, a large number of thalamic neurons will fire and provide input to a large number of cortical neurons. However, only those cortical neurons receiving the most numerous and synchronously active thalamocortical inputs would reach threshold and fire in response to activation of the "appropriate" vibrissa, thereby creating a single-whisker unit. This interpretation is consistent with our proposition that, although all cortical neurons receive thalamocortical input, the receptive field properties of cortical neurons are determined by the relative proportion of inputs they receive from specific thalamocortical and from other afferents.

Evidence for the X/Y afferent system supports the existence of segregated, parallel pathways that originate in the retina and reach the cortex—via the LGN—with little or no overlap. However, it is not known whether these parallel pathways are involved with processing different aspects of visual perception. Hubel and Livingstone (1987; Livingstone and Hubel, 1987b) suggested that, in the primate visual system, different signals involved in visual perception, for example, form, color, movement, and depth, are relayed and analyzed within separate, parallel channels.

According to Hubel and Livingstone's model, the segregation of visual

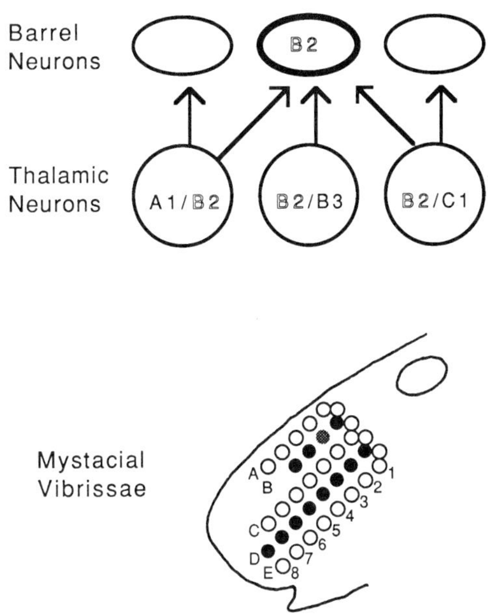

Figure 5.1. *Hypothesized circuitry for the generation of single-whisker receptive fields in the rat barrel cortex. Whiskers are designated by dots, letters designate rows of whiskers, and numbers refer to individual whiskers in each row. Stimulation of the B2 whisker results in activation of several thalamic neurons that provide input to a number of neurons in the barrel cortex. However, only those barrel neurons receiving the most numerous and synchronously active thalamocortical inputs would respond to activation of the B2 vibrrisa (after Simons et al., 1989).*

channels begins at the LGN (Figure 5.2). Cells in the parvocellular layers of the LGN usually have color-opponent receptive fields, consisting of antagonistic subregions that respond to stimuli of different wavelengths (e.g., Wiesel and Hubel, 1966). In addition, ganglion cells in the parvocellular layers are less sensitive to luminance contrast, and have smaller receptive field centers, lower conduction velocities, and more sustained responses than cells in the magnocellular layers (e.g., Derrington and Lennie, 1984). Thus, parvocellular neurons appear to be involved preferentially with the perception of color; they have a higher spatial resolution and a slower temporal resolution than that of the magnocellular neurons, which appear to be involved mainly with the perception of contrast. Some of the physiological differences between the parvo- and magnocellular ganglion cells are already present in so-called A and B retinal cells, which project to the magno- and parvocellular layers, respectively (Leventhal et al., 1981).

Cortical projections of the different classes of geniculate cells have been examined mainly with the light microscope, which cannot provide direct evidence for synaptic connections (see "Summary of Chapter 2"). Nevertheless, indirect evidence suggests that cells in the parvocellular layers provide input to cells in layer IVCβ of area 17, which in turn project to layers II/III of the cortex, specifically to regions between cytochrome oxidase-rich blobs (interblobs—see Chapter 4) (Livingstone and Hubel 1987b; Lund, 1988). Cells in layer IVCβ retain most of the receptive field properties exhibited by the parvocellular cells, but although they are color coded, most cells in layer IVCβ lack color opponency (Blasdel and Fitzpatrick, 1984). Cells in the interblob regions are mostly nonselective for stimulus color, are highly orientation selective, and respond preferentially to moving stimuli (Livingstone and Hubel, 1984b). Reciprocal pathways connect the interblob regions of area 17 with pale cytochrome oxidase stripes in area 18 (Livingstone and Hubel, 1984b). Cells in the pale stripes have receptive field properties resembling those of interblob cells, except that a high proportion of them are, in addition, end stopped (hypercomplex) (Hubel and Livingstone, 1987).

The source of inputs to cells in the blob regions of area 17 is not known, although it has been suggested that they receive input from both magno- and parvocellular layers, as well as from the interlaminar regions of the LGN (see Livingstone and Hubel, 1987b; Lund, 1988). Michael (1987) has shown that the axons of spiny stellate cells of layer

▷

Figure 5.2. *Schematic diagram of parallel channels of information processing in the primate visual system. Arrows indicate anatomical pathways between groups of cells having similar receptive field properties. Some of these pathways may be reciprocal. See the text for additional details. (After Livingstone and Hubel, 1987.)*

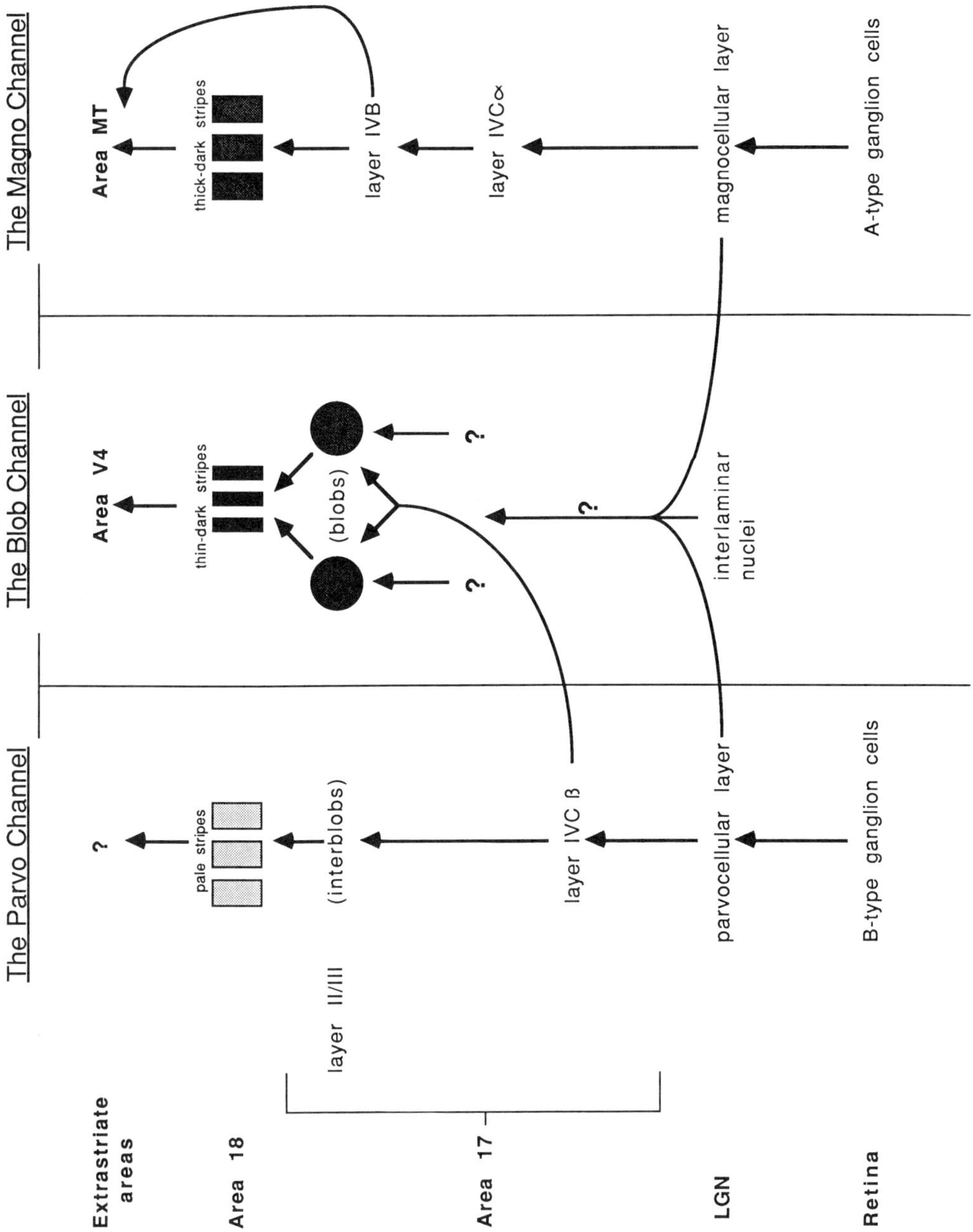

The Parvo Channel
The Blob Channel
The Magno Channel
Extrastriate areas
Area 18
Area 17
LGN
Retina
layer II/III
?
pale stripes
(interblobs)
layer IVC ß
parvocellular layer
B-type ganglion cells
Area V4
thin-dark stripes
(blobs)
?
?
?
interlaminar nuclei
Area MT
thick-dark stripes
layer IVB
layer IVCα
magnocellular layer
A-type ganglion cells

IVCβ project into the blob regions, and has suggested that this input influences the color-coded receptive field properties of the blob cells. Blob cells are not orientation selective, and about half are color opponent, whereas the rest probably convey information about brightness (Livingstone and Hubel, 1984b).

Reciprocal pathways connect the blob regions of area 17 with thin, dark cytochrome oxidase stripes in area 18 (Livingstone and Hubel, 1984b). The receptive field properties of cells in the thin stripes resemble those of blob cells (Hubel and Livingstone, 1987). Cells in the thin stripes project to area V4 (Shipp and Zeki, 1985; DeYoe and Van Essen, 1985), which is thought to be involved in pattern and color discrimination (e.g., Heywood and Cowey, 1987; Van Essen, 1985).

Cells in the magnocellular layer of the LGN project to layer IVCα of area 17, which in turn projects to layer IVB (see Livingstone and Hubel, 1987b; Lund, 1988). Layer IVB projects to the thick, dark cytochrome oxidase stripes in area 18 (Livingstone and Hubel, 1987a) and to the midtemporal cortical area (area MT; Lund et al., 1975), which also receives input from the thick stripes in area 18 (Shipp and Zeki, 1985). The receptive field properties of cells at different stages along the pathway from the magnocellular layer of the LGN are strikingly similar: At all stages, neurons are highly orientation selective, many cells in layer IVB of area 17 and in area MT are also highly direction sensitive, and many cells in layer IVB, in the thick stripes of area 18, and in area MT are involved in analyzing stereoscopic depth; a high proportion of these cells are precisely tuned for binocular disparity (Livingstone and Hubel, 1987b; Van Essen, 1985).

Based on these data and on evidence from psychophysical and clinical studies in humans, Livingstone and Hubel (1987b) have proposed the existence of three separate, parallel pathways in the primate visual system, each pathway responsible for transmitting and analyzing particular aspects of visual perception (Figure 5.2). The blob system appears to be involved preferentially with color perception, whereas the magno system seems to be involved with analyzing depth and movement. Form perception is carried by both the magno- and the parvo interblob systems, the former integrating information from different parts of the visual field, whereas the latter deals with the finer analysis of form. Although it has been suggested that the magnocellular and parvocellular channels correspond to the Y and X pathways (e.g., Enroth-Cugell and Robson, 1966), it appears that, at least in the primate visual system, these pathways are independent of one another (see Livingstone and Hubel, 1987b, p. 3418).

Both anatomical and physiological data indicate that the segregation of visual pathways is not absolute. As Livingstone and Hubel (1987b) point out, cells at any stage of these pathways may receive inputs from cells belonging to a different pathway, as well as from other cortical and subcortical regions. For example, cells in area 17 may receive inputs from both the magno- and parvocellular layers of the LGN (Malpeli et al., 1981). Divergent and convergent anatomical pathways provide additional cross-connections between the different channels (see DeYoe and Van Essen, 1988). Furthermore, different types of receptive field properties (e.g., direction and wavelength selectivities) characterize cells in more than one channel, suggesting the existence of a redundancy in processing of different aspects of visual information (cf. DeYoe and Van Essen, 1988).

These findings suggest that certain cells may belong to more than one functional pathway. The likelihood that neurons are commonly recruited for more than one circuit or pathway is discussed in detail in Chapter 7. That cells may belong to more than one channel does not refute the validity of the parallel model of visual information processing. These findings imply that the segregation of computational roles within separate channels may not be complete and that complex interactions exist between the different channels.

The parallel processing of information by spatially segregated pathways also has been identified in the somatosensory system. For example, Mountcastle et al. (1969) identified two groups of rapidly adapting neurons in the somatosensory cortex that are involved in sensing flutter vibration. Each group responds preferentially to different ranges of vibration frequencies and is activated by stimulation of different types of peripheral receptors (see also Ferrington and Rowe, 1980). Furthermore, each relay nucleus along the ascending sensory pathway to the cortex contains segregated populations of neurons that correspond to each of the two classes of "flutter-vibration" neurons in the cortex (see the review by Dykes, 1983). For a description of additional, parallel pathways of information processing in the somatosensory system, the reader is referred to the comprehensive review by Dykes (1983).

Intrinsic Pathways

Excitatory Interactions

Local synaptic connections between neurons separated by as little as a few tenths of a micrometer have been studied by cross-correlation analyses of neurons in the visual cortex (e.g., Toyama et al., 1981a). Several studies have determined that pairs of closely spaced neurons are likely to be involved in monosynaptic excitatory interactions and that the probability of the occurrence of these interactions decreases as the spatial separation between the pair of neurons increases (Abeles, 1982; Michalski et al., 1983; Ts'o et al., 1986). The powerful excitatory interactions between closely spaced cells can be related to the elaborate ramifications of axon collaterals of neurons in the vicinity of their cell bodies (see Chapter 2) and to the concept expressed in the corollary to Rule 1 (Chapter 3) that an axonal pathway will form synapses with every element in its target region capable of forming the type of synapse normally made by the pathway. Intrinsic excitatory interactions may involve pyramidal neurons or nonpyramidal neurons, such as spiny stellate cells, some bipolar neurons, and certain double bouquet cells, which form asymmetric, presumably excitatory synapses (see Table 3.1). In addition to their local branches, axon collaterals may extend over long distances horizontal to the pial surface.

Excitatory monosynaptic interactions between cells in the visual cortex of the cat may extend horizontally for distances of up to 3 mm (e.g., Ts'o et al., 1986). Because these interactions are excitatory (Toyama et al., 1981a, b; Ts'o et al., 1986), the most likely anatomical substrates for them are the local axon collaterals of pyramidal neurons, for these cells have axons that may travel for several millimeters within the cortex (see Chapter 2). Excitatory interactions in the visual cortex occur among cells residing within columns of like orientation selectivity, but not among neurons within columns of different orientation selectivities (Michalski et al., 1983; Toyama et al., 1981b; T'so et al., 1986). In the monkey primary visual cortex, intrinsic connections occur among neurons in cytochrome oxidase-rich blobs that contain cells having color-specific and nonorientation-selective receptive field properties and among interblob regions that contain cells selective for stimulus orientation but not for color (Livingstone and Hubel, 1984a). In area 18 of the cat, intrinsic excitatory interactions have also been shown to be specific, in that they occur preferentially between neurons whose receptive field orientation selectivities are orthogonal to one another (Matsubara et al., 1985, 1987). Excitatory connections link regions of the cat auditory cortex that contain neurons having similar frequency preferences (Matsubara and Phillips, 1988). Excitatory interactions between pairs of neurons sharing similar receptive field properties have also been identified in the somatosensory cortex of the cat, where cross-correlation analyses have revealed short latency excitatory interactions between pairs of neurons belonging to the same submodality band (Metherate and Dykes, 1985).

Ts'o et al. (1986) suggested that the simultaneous activation of neurons within two columns are mediated not only by direct connections between the pair of cells, but also, and perhaps more often, by common input these cells receive from neurons in other columns of like specificity. In this way, thalamic input to a single orientation column could be transmitted, by feed-forward excitation, to other columns containing cells that share similar response properties.

Synaptic interactions between laminae have been studied much more extensively than horizontal synaptic interactions. Toyama et al. (1974) examined interlaminar synaptic interactions in the cat visual cortex following stimulation of extrinsic afferents. Cells in layers III to V showed monosynaptic excitatory postsynaptic potentials (EPSPs) following LGN stimulation, whereas cells in layers II and VI showed disynaptic EPSPs, generated in a feed-forward excitation mechanism. Slightly different findings were reported by Ferster and Lindström (1983), who identified units in layers III, IV, upper V, and VI of the cat visual cortex that were monosynaptically activated by LGN afferents, whereas units in layers II, III, V, and VI were di- or trisynaptically activated by LGN afferents.

Circuitry responsible for the disynaptic excitation of cells outside layer IV has been interpreted to involve axon collaterals belonging to pyramidal neurons, because the axonal arborizations of spiny stellate cells ramify only within layer IV or immediately above or below this layer (see Ferster and Lindstöm, 1983). However, because many types of cells having somata in layers III to VI receive direct thalamic input (see Tables 3.2 and 5.1), the identification of cells providing feed-forward excitation becomes complicated. Thus, any neuron in cortical layers III to VI may be responsible for the disynaptic excitation of cortical neurons, following LGN stimulation.

Hubel and Wiesel's (e.g., 1977) model of hierarchical processing uses interlaminar synaptic connectivity as a basis for the sequential activation of cells having progressively more 'complex' receptive fields. We have already seen that the simple/complex dichotomy is not correlated with laminar organization or with specific morphological types of neurons. However, as mentioned above, hierarchical processing may still proceed from 'simple' to more 'complex' cells, irrespective of their laminar location or morphological characteristics. This possibility can be evaluated by determining the intracortical interactions between cells whose receptive field properties have been determined.

One approach to determining the contribution of intracortical interactions for shaping the receptive field properties of cortical neurons is to investigate the effects on cortical neurons of the selective inactivation of specific intracortical inputs. This approach was followed by Malpeli (1983; Malpeli et al., 1986), who showed that in the absence of thalamic input simple cells in layers IV and VI of the visual cortex are silenced, but that complex cells in the superficial layers retain their receptive field properties. These findings imply that the activity of cells with complex receptive fields are not dependent on input from cells having simple receptive fields. Moreover, simple cells respond preferentially to slow stimulus velocities, whereas complex cells respond preferentially to significantly faster stimulus velocities (e.g., Pettigrew et al., 1968; Movshon, 1974). This indicates that simple cells are unlikely to provide direct input to cells with complex receptive field properties. In addition, activation of local axon collaterals belonging to layer VI corticogeniculate neurons, some of which have complex receptive field properties (e.g., Gilbert, 1977; Harvey, 1980), produces monosynaptic EPSPs in simple cells in layer IV (Ferster and Lindström, 1985), suggesting that complex cells may provide input to simple cells. This assumption is supported by studies showing that the activity profile of simple cells may be modulated by input from cells having complex receptive field properties (e.g., Burr et al., 1981; Maffei, 1985).

The hierarchical model suggests that inputs from supragranular layers to the infragranular layers are responsible for the formation of the receptive field properties of neurons in the infragranular layers. However, inactivation of cells in the supragranular layers, either by cooling or lesioning these layers, results in the disarrangement of the receptive field properties of only the "special complex" cells in layer V; the receptive field properties of other cells in layers V and VI remain intact (Schwark et al., 1986).

A direct approach to studying intracortical interactions involves using cross-correlation analyses to determine the types and strengths of interactions between cells whose receptive field properties have been defined. Using this approach, several studies have shown that intracortical monosynaptic excitatory interactions occur almost exclusively between cells with complex receptive properties, and then only when the pair of cells display similar orientation selectivities (e.g., Michalski et al., 1983; Toyama et al., 1981b). Excitatory interactions from cells with simple receptive fields to cells with complex receptive fields have not been observed in these studies.

In conclusion, the results of these studies do not support the assumption included in the hierarchical model that the receptive field properties of complex cells are the result of excitatory inputs from simple cells. Simple cells may provide input to complex cells, but the available data suggest that the influence of this input is less important for shaping the receptive field properties of complex cells than the hierarchical model would have us believe.

Intracortical circuits involving neurons within different cortical laminae, and within different functional columns, provide excitatory links between neurons that share similar receptive field properties. However, excitatory interactions alone cannot provide a complete description of the cortical circuitry involved in the formation of the receptive field properties of cortical neurons. For a complete picture of cortical circuitry, it is also necessary to consider the role of inhibitory interactions in the formation of the receptive field properties of cortical neurons.

Inhibitory Interactions

Electrical stimulation of cortical afferents does not induce monosynaptic inhibitory postsynaptic potentials (IPSPs) in the cerebral cortex (e.g., Ferster and Linsdtröm, 1983; Toyama et al., 1974), and so inhibitory interactions are presumed to be mediated only by intracortical mechanisms. In addition, afferents to the cerebral cortex form only asymmetrical synapses, which are presumed to be excitatory; symmetrical, presumed inhibitory synapses are of intracortical origin only (see Chapter 3). Many, if not all, inhibitory interactions in the cerebral cortex may be mediated by GABA, since no other inhibitory transmitter, or putative neurotransmitter, has been identified with certainty in the cerebral cortex (e.g., Krnjević, 1984). In addition, immunocytochemical studies have demonstrated that terminals forming symmetrical synapses are almost exclusively GABAergic (Keller and White, 1987, Table 1).

Cross-correlation analyses have revealed horizontal inhibitory interactions in the visual cortex of the cat (e.g,. Michalski et al., 1983; Toyama et al., 1981a, b), but it is not known if these inhibitory interactions resemble their excitatory counterparts by linking neurons with similar receptive field properties. Evidence that this may be true has been provided in a study of the cat somatosensory cortex, in which inhibitory interactions were shown to link neurons of like submodalities (Martin and Spencer, 1985). Nelson and Frost (1978) have suggested that inhibitory interactions in the visual cortex link neurons sharing similar optimal orientations and that these interactions serve to sharpen the orientation selectivities of these cells. The horizontal expanse of inhibitory interactions is more restricted than that of the excitatory ones, a finding that is in agreement with anatomical studies of the cat visual cortex showing that the horizontal spread of axons belonging to basket cells, a likely candidate for inhibitory interactions (see Chapters 3 and 6), rarely reaches 1 mm (Somogyi et al., 1983a), whereas pyramidal cell axons may extend horizontally for up to 8 mm (Gilbert and Wiesel, 1979).

Horizontal, inhibitory interactions may be involved in lateral inhibition, as first proposed by Mountcastle (1957). According to Mountcastle's model for two-point discrimination, two stimuli applied to separate regions of the skin set up two excitatory gradients at every relay point in the somatic sensory system, including the cortex. Each excitatory region is flanked by an inhibitory surround that sharpens the excitatory peak and further enhances the contrast between the two excitatory peaks. In the barrel cortex of the rat, fast-spiking units, which are thought to be GABAergic neurons, commonly receive convergent, multiwhisker inputs (Simons, 1978). These cells, in a process of feed-forward inhibition, would lower the probability that other cortical neurons would fire. In addition, lateral inhibition may occur between functional columns; the circuitry required for these interactions would involve direct thalamocortical input to inhibitory neurons, which would then interact with neurons situated in another modality-specific column.

Evidence for such a circuit was provided for the mouse barrel cortex (Keller and White, 1987), where it was shown that a GABAergic neuron in one barrel (barrel A) had one of its dendrites in an adjacent barrel (barrel B), where its dendrite received thalamocortical synapses (Figure 5.3). Because neurons in a single barrel are activated preferentially by input from a single vibrissa (e.g., Simons, 1978), thalamic input to barrel B could excite the GABAergic neuron, which, in turn, would inhibit the activity of neurons in barrel A, which contains its cell body and axonal ramification. Barrel A receives input from a different vibrissa, and so inhibition of neurons within it would presumably increase the discrimination between inputs from the adjacent vibrissae that correspond to barrels A and B. Indeed, it has been shown that some neurons in the barrel cortex of the rat are inhibited when two adjacent vibrissae are stimulated (Simons, 1983) and that these inhibitory interactions between adjacent whiskers are mediated by intracortical mechanisms (Carvell and Simons, 1987; Simons and Carvell, 1989). Similar findings were observed in the somatosensory cortex of monkeys, where it was concluded that activation of cells in one column was accompanied by inhibition of cells in neighboring columns (Mountcastle and Powell, 1959).

Local injections of tritiated-GABA, which is thought to be transported retrogradely within GABAergic neurons, label cell bodies in the monkey sensorimotor and visual cortices for a vertical extent of approximately 1.5 mm (e.g., DeFelipe and Jones 1985). In comparison, the horizontal distribution of labeled cell bodies is restricted to from 600 to 700 μm from the injection site, indicating that the inhibitory horizontal interactions are not as extensive as vertical interactions (DeFelipe and Jones, 1985; Somogyi et al., 1981, 1983b).

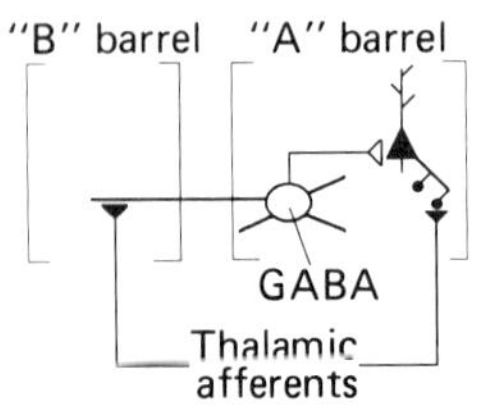

Figure 5.3. *Synaptic circuitry that may form an anatomical substrate for lateral inhibition in the mouse barrel cortex. Blackened neurons designate cells that form excitatory synapses; unfilled neurons form inhibitory synapses. GABA = GABAergic neuron.*

These studies also show that interactions between GABAergic cells in the superficial and deep layers are reciprocal. The application of GABA to neurons in layer VI of the visual cortex inactivates these neurons and, in addition, abolishes end inhibition in cells of the superficial layers (Bolz and Gilbert, 1986). Bolz and Gilbert (1986) suggested that their findings can be explained by inactivation of excitatory projections from layer VI to the superficial layers, but their data are also compatible with a suppression of inhibitory interactions from layer VI to the superficial layers. Moreover, GABAergic neurons in layer VI of the visual cortex of the cat (Kisvarday et al., 1987) and monkey (DeFelipe and Jones., 1985) have extensive projections to superficial layers. These findings demonstrate the existence of extensive vertical inhibitory interactions that link neurons within a functional column, and suggest a pivotal role for this interaction in shaping the receptive field properties of cortical neurons.

Synaptic interactions in the cerebral cortex provide mechanisms for both feed-forward and feedback inhibition. Electrical stimulation of thalamocortical afferents to the visual cortex of the cat produces monosynaptic EPSPs and disynaptic IPSPs in cells in layers III to VI (e.g., Ferster and Lindström, 1983), and some of these cells have been shown to be corticofugal neurons (Toyama et al., 1974). Similarly, in the turtle visual cortex, photic stimulation activates both pyramidal and nonpyramidal cells, the latter inhibiting the former in a process of feed-forward inhibition (Kriegstein, 1987). In the mouse barrel cortex, electrical stimulation of the thalamus produces monosynaptic EPSPs and disynaptic IPSPs in neurons of layers II through VI (Agmon and Connors, 1987).

These interactions suggest cortical circuits in which thalamocortical input excites both pyramidal neurons and GABAergic cells. The GABAergic cells, via feed-forward inhibition, inhibit the pyramids that project to other cortical or subcortical areas. A similar circuit, involving interactions between thalamocortical afferents, pyramidal tract neurons, and inhibitory cells, has been identified by extracellular recordings of neurons in the monkey motor cortex (Steriade and Deshênes, 1973). Blum (1974) has provided evidence for synaptic interactions in the sensorimotor cortex of the cat and monkey, in which axon collaterals excite inhibitory neurons, which then provide feed-forward inhibition to other pyramidal neurons. Pertinent to this are findings from a study of the mouse somatosensory cortex in which thalamic afferents were observed to form synapses with both GABAergic neurons and corticothalamic projection cells (Keller and White, 1989). This arrangement would provide for an excitatory pathway from corticothalamic cells to GABAergic neurons and for an inhibitory pathway from GABAergic neurons to corticothalamic cells.

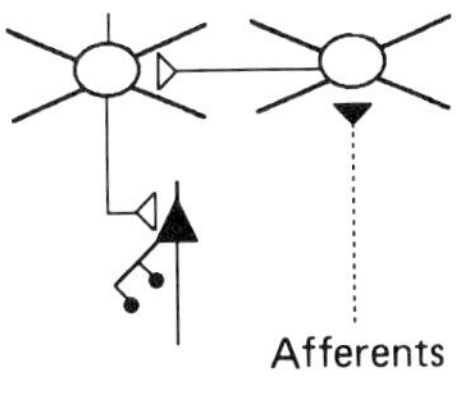

Figure 5.4. *Synaptic circuitry involved in disinhibition. Blackened neurons designate cells that form excitatory synapses; unfilled neurons form inhibitory synapses.*

In layer IV of the cat visual cortex, most neurons show monosynaptic EPSPs and disynaptic IPSPs (Ferster and Lindström, 1983; Toyama et al., 1974), following thalamic stimulation. These interactions imply that thalamic afferents activate layer IV spiny stellate cells and GABAergic neurons, the latter exerting feed-forward inhibition onto the spiny stellate cells. Alternatively, the circuitry may involve thalamocortical input to two GABAergic neurons, one of them inhibiting the other. Disinhibition of inhibitory neurons (Figure 5.4), following the activation of thalamocortical afferents, has also been observed in the pericruciate cortex of the cat (Kelly and Renaud, 1974). Because non-

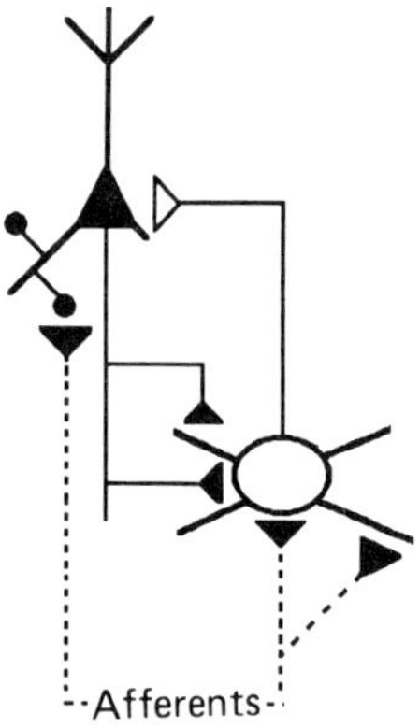

Figure 5.5. *Triad. Blackened neurons designate cells that form excitatory synapses; unfilled neurons form inhibitory synapses.*

pyramidal, presumably GABAergic, neurons are distinguished by their lack of adaptation (e.g., McCormick et al., 1985), these cells would fire continously in response to tonic stimuli and thus exert a tonic inhibition on their postsynaptic targets. Disinhibitory interactions could remove this tonic inhibition from cortical neurons, thereby altering their receptive field properties (see also Chapter 7).

Other cells in the cat visual cortex, mainly in layers II and VI, show disynaptic EPSPs and trisynaptic IPSPs following activation of LGN afferents (Ferster and Lindstöm, 1983; Toyama et al., 1974). Similarly, electrical activation of the thalamus produces disynaptic EPSPs and trisynaptic IPSPs in neurons in layers II to III of the mouse barrel cortex (Agmon and Connors, 1987). Several types of cortical circuits can be inferred from these interactions, all of which involve feed-forward excitation from an interneuron to excitatory and/or inhibitory cells. Each of these patterns of circuitry shares a basic principle: common input to both inhibitory and excitatory cells, and reciprocal synaptic interactions between these cells. This type of arrangement has been termed a *synaptic triad* (Figure 5.5). The realization that many of the synaptic interactions revealed by electrophysiological methods can be described in terms of triadic arrangements lends additional support to the proposition that triadic arrangements are a basic element of cortical circuitry (see Chapter 7, "Synaptic Triads").

Summary of Chapter 5

Studies employing electrophysiological techniques have revealed patterns of synaptic interactions in the cerebral cortex that confirm and extend the results obtained by neuroanatomical methods. Of the extrinsic afferent inputs to the cerebral cortex, the thalamocortical projections have been studied most thoroughly. A wide variety of neurons, having different morphological and receptive field properties, is postsynaptic to thalamocortical afferents. In the visual system, different types of thalamocortical afferents are involved in segregated, parallel pathways of information processing. These parallel pathways are involved in transmitting and processing different aspects of visual perception, such as color, movement, and depth.

Intrinsic cortical connections, both excitatory and inhibitory, play a major role in synaptic interactions within the cerebral cortex. Both excitatory and inhibitory interactions form highly specific connections within and between functional columns. These intrinsic interactions are responsible for shaping the receptive field properties of cortical neurons and for the formation of functional columns. A common pattern of circuitry revealed is that of the triad, involving common input to both inhibitory and excitatory neurons, and reciprocal synaptic interactions between these neurons.

The cortical circuitry described thus far in Parts I and II of this book have centered on the elucidation of synaptic interactions and on how these interactions may determine the receptive field properties of cortical neurons. The rapidly accumulating data on the functional role of GABA in shaping the receptive field properties of cortical neurons warrant particular consideration of the mechanisms of GABAergic inhibition, and these are discussed in Chapter 6.

6 GABAergic Inhibition in the Cerebral Cortex

GABA-Mediated Inhibition

One of the major advances in our understanding of cortical function has been the elucidation of the role of neurotransmitters and other neuroactive substances in cortical circuitry. At present, more is known of the anatomical distribution and of the physiological effects of GABA than of any other neuroactive substance. Microiontophoretic techniques, whereby minute quantities of putative transmitters, modulators, or drugs are applied to neurons, were first employed by Curtis and Eccles (1958) to study neurons in the spinal cord. This approach was subsequently employed to study the effects of GABA on cortical neurons.

Practically all cortical neurons are powerfully inhibited by local applications of GABA, and a close correlation exists between IPSPs and the membrane actions produced by application of GABA, in that both effects are mediated by an increase in chloride conductance (e.g., Dichter, 1980; Dreifuss et al., 1969; Krnjević and Schwartz, 1967). The revelation that GABA is released from cortical neurons in response to stimulation of the cortex (e.g., Iversen et al., 1971) provided further evidence that GABA is a major inhibitory neurotransmitter in the cerebral cortex. A large body of data supports the notion that GABAergic inhibition is a main determinant of the functional properties of cortical neurons (e.g., Dykes et al., 1984; Sillito, 1984; Tsumoto et al., 1979). In addition, GABA has been implicated in various pathological conditions, such as epilepsy (Giaretta et al., 1985; Ribak et al., 1979, 1986), Huntington's disease (Penny and Young, 1982), and cortical hemiplegia (Brailowsky et al., 1986). The cardinal role of GABA in normal as well as in pathological processes in the cortex has stimulated studies aimed at the identification of those neurons containing and presumably utilizing GABA as a neurotransmitter.

Anatomical Demonstration of GABAergic Neurons

Neurons utilizing GABA (GABAergic) have been identified by histochemical methods (Araki et al., 1984; Nagai et al., 1983), autoradiography (Hamos et al., 1983; Somogyi et al., 1984; Winer, 1986), and immunocytochemistry. Immunocytochemical methods have employed an antibody to GABA (Keller and White, 1987; Ottersen and Storm-Mathisen, 1984a, b; Somogyi et al., 1985), or to glutamic acid decarboxylase (GAD), the rate-limiting enzyme in the synthesis of GABA (e.g., Houser et al., 1983; Keller and White, 1986; Mugnaini and Oertel, 1985; Ribak, 1978).

It has been estimated that GABAergic neurons constitute 20% of all neurons in the cat primary visual cortex (Gabbot and Somogyi, 1986) and 25% of the neuronal population in most areas of the monkey cortex (Hendry et al., 1987). GABAergic neurons in the cerebral cortex are exclusively of the nonpyramidal cell type. GABAergic neurons that are non- or sparsely spinous nonpyramidal cells have been identified in the sensory cortex of the rat, mouse (Chmielowska et al., 1988; Keller and White, 1986, 1987; Lin et al., 1985), and monkey (Hendry et al., 1987; Houser et al., 1983); in the motor cortex of the monkey (Hendry et al., 1987; Houser et al., 1983) and mouse (Solberg et al., 1988); in the visual cortex of the rat (Ribak, 1978; Somogyi et al., 1984), cat (Bear et al., 1985; Freund et al., 1983; Hamos et al., 1983; Somogyi et al., 1985), and monkey (Fitzpatrick et al., 1987; Hendry et al., 1987); in the cat primary auditory cortex (Winer, 1986); in the rat prefrontal and frontoparietal cortices (Esclapez et al., 1987); in the opossum piriform cortex (Haberly et al., 1987); and in the turtle dorsal cortex (Blanton et al., 1987). Some nonpyramidal neurons contain, in addition to GABA, one or more types of neuropeptide (see Chapter 2).

The types of nonpyramidal neurons containing GABA include chandelier cells, basket cells, local plexus neurons, and several types of vertically oriented neurons (see Chapter 2). It is unknown whether all neurons of these types contain GABA; the possibility exists that some of these neurons are not GABAergic. However, examination of Golgi-impregnated or HRP-filled neurons for GABA (Freund et al., 1983; (Kisvarday et al, 1986; 1987; Somogyi et al., 1985) suggests that every nonpyramidal neuron, of the classes described above, is GABAergic.

Similar findings were reported by Meinecke and Peters (1987), who used correlated light and electron microscopy to determine the types and proportions of nonpyramidal neurons containing GABA. These authors report that multipolar, bitufted, bipolar, and horizontal cells in the rat visual cortex are immunoreactive for GABA. In addition, by examining 950 neurons in serial thin sections, it was determined that, except for 20% of the bipolar cell population, each of the other types of nonspiny, nonpyramidal neurons in the tissue contained GABA.

Physiological Basis of GABA-Mediated Inhibition

It is generally agreed that GABA inhibits the generation of spikes in cortical neurons by generating hyperpolarizing inhibitory postsynaptic potentials (IPSPs). IPSPs prevent the membrane potential of the initial axon segment from reaching threshold for spike generation (e.g., Iversen et al., 1971; Krnjević and Schwartz, 1967). GABA-mediated effects may also inhibit neurons by a process termed *shunting* in which the GABA-mediated synaptic activity opens chloride channels, thereby increasing membrane conductance and reducing the amplitude of postsynaptic potentials produced by excitatory inputs to the cells (e.g., Koch et al., 1982; Kriegstein and Connors, 1986). Shunting effects, produced by GABAergic input to a dendrite, are capable of controlling and modulating inputs at more distal portions of the dendritic tree (e.g., Koch et al., 1982), and thus, these effects may have a role in shaping the receptive field properties of cortical neurons.

Segev and Parnas (1983) have proposed that the effects of IPSPs on EPSPs are determined by the temporal delay between the appearance of the two synaptic events and by the nature of the IPSP. Thus, a hyperpolarizing IPSP could reduce the size of an EPSP, whereas a depolarizing IPSP could enhance the amplitude and lengthen the time integral of the EPSP (Segev and Parnas, 1983).

Local application of GABA in the vicinity of the cell bodies of cortical neurons produces hyperpolarizing responses, whereas application of GABA close to the cells' dendrites produces depolarizing responses. Both of these effects are associated with an increase in conductance and with inhibition in the postsynaptic cell, but the depolarizing response may also be excitatory (Kemp, 1983; Kriegstein and Connors, 1986; Scharfman and Sarvey, 1985, 1987). Both the somatic and dendritic responses produced by the application of GABA can be recorded from pyramidal and nonyramidal neurons in any layer of the cortex (Scharfman and Sarvey, 1987). Similar effects are produced by local applications of GABA to cell bodies and dendrites of hippocampal neurons (Andersen et al., 1980; Misgeld et al., 1986). Misgeld et al. (1986) have suggested that the GABA-mediated depolarizing and hyperpolarizing IPSPs in hippocampal neurons are both generated by chloride transport processes, one that accumulates chloride and one that extrudes chloride.

These findings suggest that GABAergic terminals may exert different effects on different postsynaptic elements. Inhibitory input to the cell body will interact with other synaptic inputs, and the balance between the EPSPs and IPSPs will determine whether the membrane potential will reach the threshold for action-potential generation. In contrast to the axosomatic influences, GABAergic axodendritic synapses may produce a shunting effect, which may either facilitate or inhibit the input from more remotely placed excitatory synapses. It is interesting to note that GABAergic synapses and unlabeled, symmetrical synapses are distributed preferentially on the proximal portions of dendrites, on somata, and in the vicinity of dendritic branch points (e.g., Keller and White, 1987, 1989; White and Rock, 1980; White et al., 1984). GABAergic terminals synapsing near dendritic branch points would be capable of shunting the input from one dendritic branch while facilitating the input from other branches (e.g., Koch et al., 1982). Assuming that a widely spread dendritic tree receives input from various sources at different dendritic branches, this mechanism could control the flow of inputs to the cell body and thus participate in shaping the receptive field properties of the cell. For example, it has been postulated (Emerson et al., 1987) that "shunt" interactions by axodendritic synapses are involved in determining the orientation selectivity of neurons in the visual cortex.

GABAergic terminals synapsing with axon initial segments are strategically placed to exert an exceptionally powerful inhibitory influence near the spike trigger zone of the postsynaptic cell. This interaction would be capable of preventing the propagation of action potentials, even in the presence of powerful excitatory inputs to the cell body and dendrites. Inhibition at the axon initial segment may act as a "frequency filter," controlling the rate of firing of the postsynaptic cell.

GABAergic Inhibition and Receptive Fields

The widespread distribution of GABAergic neurons in the cerebral cortex, and the fact that every type of cortical neuron receives GABAergic input (e.g., Houser et al., 1984), suggest that GABAergic interactions have a significant role in the processing of information in the cerebral cortex. The contribution of GABA-mediated inhibitory mechanisms to the response properties of cortical neurons has been evaluated by blocking the effects of GABA with specific pharmacological agents and comparing the functional properties of cortical neurons before and after the induction of GABA antagonism (e.g., Pettigrew and Daniels, 1973; and see review by Sillito, 1984). The most commonly employed neuropharmacological technique involves local injections of bicuculline, an effective GABA antagonist, and electrophysiological recordings from neurons in the vicinity of the injection site (e.g., Rose and Blackmore, 1974).

Application of bicuculline has been shown to affect several response properties of cortical neurons, suggesting a significant role for GABAergic inhibition in the shaping of these response properties. Not all neurons are affected to the same extent by local applications of bicuculline. Variations in the responses to bicuculline may be due to an incomplete antagonism of the GABAergic effects (see Sillito, 1984). However, the possibility that the receptive field properties of different cells are influenced to different extents by GABAergic inhibition, or that additional, non-GABAergic interactions play a large role in the formation of some receptive field properties, cannot be excluded.

The response magnitude of both simple and complex neurons in the cat visual cortex is increased, following GABA antagonism (Rose and Blackmore, 1974; Tsumoto et al., 1979; Sillito, 1975, 1977; Wolf et al., 1986). The same effect can be produced by bicuculline administration to neurons in the cat somatosensory cortex (Dykes et al., 1984; Hicks and Dykes, 1983). The increase in response magnitude can be ascribed to the removal of tonic GABA-mediated inhibition, which would otherwise suppress the activity of cortical neurons.

The discrete ON and OFF subregions in the receptive fields of simple (Sillito, 1975; Wolf et al., 1986) and certain complex (Rose and Blackmore, 1974) neurons in the cat visual cortex are lost following GABA antagonism. The loss of discrete ON and OFF subregions by suppression of GABAergic inhibition suggests that these subregions are not determined simply by specific excitatory input from the LGN to simple cells, or from simple to complex cells. Rather, these findings suggest that intracortical inhibition is responsible for keeping the ON and OFF subregions spatially separated.

It is generally believed that the receptive fields of simple cells are constructed by the convergence of inputs from ON- and OFF-center cells in the LGN (see the review by Henry, 1985). If inhibitory interactions are responsible for the spatial separation of these inputs, by selectively suppressing parts of the receptive fields of the postsynaptic cells, the receptive fields of the inhibitory cells need also be spatially organized; that is, GABAergic neurons would receive inputs from spatially separated ON- and OFF-type thalamic afferents or from only one type of afferent (cf., Sillito, 1984).

GABAergic neurons receiving direct thalamic input have been identified in the visual cortex (Hamos et al., 1983) and in other cortical areas, such as the somatosensory cortex (e.g., Keller and White, 1987). In addition, thalamocortical afferents having ON- or OFF-center receptive fields are segregated in the visual cortex of several species (e.g., LeVay et al., 1987; Norton et al., 1985), and some simple cells in the cat visual cortex may receive input from only one type of afferent (Lee et al., 1977). If some GABAergic neurons receive input from only one type of afferent, the receptive field properties of these cells would not be affected by GABA antagonism. Therefore, these cells could represent the subpopulation of neurons whose spatially organized receptive fields are not affected by bicuculline application (Pettigrew and Daniels, 1973; Sillito, 1984; Wolf et al., 1986).

Directional specificity is another receptive field feature that is lost as a result of GABA antagonism (Rose and Blackmore, 1974; Sillito, 1977, 1984; Wolf et al., 1986). The loss of directional specificity is not due to increased input from a nonpreferred direction, which would otherwise be inhibited, because directional specificity is not eliminated by application of an excitatory amino acid (Sillito, 1977). Therefore, directional specificity appears to be influenced by specific intracortical inhibitory mechanisms and not solely by excitatory afferent inputs, as suggested by the hierarchical model (cf., Goodwin et al., 1975a, b; and see Chapter 4).

Confirmatory evidence for this proposition is provided by studies showing that a stimulus moving in a nonpreferred direction induces a powerful inhibition in directionally specific cells (Creutzfeldt et al., 1974). In addition, Duysens (1987) has shown that directional specificity can arise from suppressive zones outside the discharge areas. Similarly, Emerson et al. (1987) have shown that directional specificity in the receptive field properties of complex cells is dependent on suppressive interactions elicited by stimuli moving in the null direction. Inactivation of inputs from cortical regions representing motion in the nonpreferred direction results in a loss of directional selectivity of cortical neurons, indicating that direction selectivity involves inhibitory intracortical inputs (Eysel et al., 1987).

The orientation selectivity of both simple and complex neurons in the cat visual cortex is markedly affected by the application of bicuculline (Pettigrew and Daniels, 1973; Rose and Blackmore, 1974; Sillito, 1975, 1984; Tsumoto et al., 1979; Wolf, 1986). The loss of orientation selectivity following antagonism of GABA-mediated inhibition runs counter to the suggestion that orientation selectivity of neurons in the visual cortex is determined solely by the organization of excitatory inputs (e.g., Hubel and Wiesel 1962). Swadlow and Weyand (1987) have shown that neurons in the rabbit visual cortex that respond to thalamic input that is presumed to be monosynaptic are not orientation selective, whereas neurons activated polysynaptically by thalamic stimulation are orientation selective.

Approximately half of the complex cell population, as well as some simple cells, show a reduction but not a loss of orientation selectivity following GABA antagonism (Sillito, 1975, 1984; Tsumoto et al.,

1979), suggesting that GABAergic inhibition is not the only mechanism shaping the orientation selectivities of cortical neurons. Schall et al. (1986) have shown that the distribution of preferred orientation of cortical neurons matches the orientations of the retinal ganglion cells subserving the same region of the visual field. Thus, it appears that, at least for some neurons, orientation selectivity is specified by inputs to the cortex; in these instances, the role of GABAergic inhibition is to enhance and refine orientation selectivity (see also Ferster and Koch, 1987).

This conclusion is supported by studies that have established the cardinal role of intracortical, inhibitory interactions in the formation of orientation-specific receptive fields (e.g., Nelson and Frost, 1978). For example, the firing rates of cells in the cat visual cortex are attentuated by a visual stimulus having an orientation other than that preferred by the cell (Burr et al., 1981; Creutzfeldt et al., 1974; Ramoa et al., 1986). Cross-orientation inhibition in the cat visual cortex has been shown to be mediated by GABAergic inhibition (e.g., Morrone et al., 1987).

An additional receptive field property of cortical neurons, namely, their preferential response to input from one eye (ocular dominance), is also affected by GABA antagonism. Sillito et al. (1980) have shown that the ocular dominance properties of aproximately 50% of the neurons in the cat visual cortex are affected by the local application of bicuculline. These findings suggest that thalamic input from both eyes terminates within a single ocular dominance column and that intracortical, GABAergic processes are involved in the suppression of inputs originating from one of the eyes (Sillito, 1984).

Several studies have provided evidence that these GABAergic interactions are an example of lateral inhibition between adjacent ocular dominance columns. For example, Tsumoto (1978) has shown that cortical neurons driven exclusively or preferentially by inputs from one eye are inhibited by stimulating the contralateral optic nerve.

The results of studies employing bicuculline administration to neurons in the cat somatosensory cortex have been reviewed by Hicks et al. (1985). These studies have shown that the specificity of neurons to activation by inputs from appropriate modality and submodality channels is not affected by GABA antagonism, indicating that GABA-mediated inhibition is not involved in the segregation of submodalities in the somatosensory cortex (Dykes et al., 1984; Hicks and Dykes, 1983). However, the administration of bicuculline does result in an increase in the receptive field sizes of rapidly adapting neurons; most slowly adapting neurons are unaffected by GABA antagonism (Dykes et al., 1984; Hicks and Dykes, 1983; Hicks et al., 1985).

Gardner and Costanzo (1980b) have suggested that intracortical inhibition is responsible for the property of directional specificity in somatosensory neurons. They proposed that lateral inhibition is the most likely mechanism underlying these interactions. Similarly, Simons (1985) concluded that intracortical inhibition is responsible for the directional specificity of neurons in the rat barrel cortex. In the whisker projection area of the cat cortex, intracortical inhibitory mechanisms have been shown to play a major role in the transformation of the receptive field properties of thalamocortical afferents (Hellweg et al., 1977).

Summary of Chapter 6

Neurons containing the inhibitory neurotransmitter GABA form a significant proportion of the neuronal population in the cerebral cortex of every species. Available evidence suggests that every cortical neuron is subject to inhibition mediated by GABAergic synapses. GABAergic synapses are preferentially distributed along strategic points on the neuronal membrane, where they exert specific influences on the activities of the cell.

GABA-mediated inhibition plays a pivotal role in shaping the receptive field properties of neurons in the cerebral cortex: Several characteristics of the receptive field properties of cortical neurons are completely abolished, or significantly altered, following the removal of GABAergic inhibition. Therefore, GABA-mediated inhibition must be considered in any evaluation of models of information processing in the cerebral cortex. In Part III several such models will be reviewed and evaluated, based on the data presented in Parts I and II of this book.

General Aspects of Brain Circuitry (Continued)

The following generalizations, based on findings presented in Part II, are offered as rules for the shaping of receptive field properties of cortical neurons. These assumptions, and the complementary rules offered in Part I, form the basis of the critical evaluation of models of cortical information processing presented in Part III.

Rule 1. The receptive field properties of every cortical neuron are shaped by the spatial and temporal integration of inputs from a variety of excitatory and inhibitory sources. Inputs from a single source cannot be the sole determinants of the receptive field properties of cortical neurons.

Rule 2. Only a fraction of the synaptic inputs to a cortical neuron are activated at one time. Therefore, the receptive field properties of cortical neurons are transitory and are determined by the cortical circuitry active at a given time.

Rule 3. Excitatory and inhibitory synaptic interactions between cortical neurons preferentially link neurons situated in close proximity to one another, and these interactions typically link neurons having similar receptive field properties. Synaptic interactions between closely spaced neurons, having similar receptive field properties, provide a basis for the similarity of receptive field properties of neurons within a functional column.

References for Part II

Abeles, M. (1982). *Local Cortical Circuits: An Electrophysiological Study*. Springer-Verlag, Berlin.

Abeles, M., and Goldstein, M.H., Jr. (1970). Functional architecture in cat primary auditory cortex: Columnar organization and organization according to depth. *J. Neurophysiol. 33*, 172–187.

Adrian, E.D., and Morruzi, G. (1939). Impulses in the pyramidal tract. *J. Physiol. 97*, 153–199.

Agmon, A., and Connors, B.W. (1987). Membrane properties and thalamocortical synaptic responses in mouse barrel cortex neurons: An intracellular study in vitro. *Soc. Neurosci. Abst. 13*, 73.3.

Albus, K. (1979). ^{14}C-Deoxyglucose mapping of orientation subunits in the cat's visual cortical areas. *Exp. Brain Res. 37*, 609–613.

Albus, K., and Donate-Oliver, F. (1977). Cells of origin of the occipitopontine projection in the cat: Functional properties and intracortical location. *Exp. Brain Res. 28*, 167–174.

Alloway, K.D., and Burton, H. (1985). Submodality and columnar organization of the second somatic sensory area in cats. *Exp. Brain Res. 61*, 128–140.

Amassian, V.E. (1953). Single unit analysis of activity of cortical neurons. *Proc. 19th Int. Congr. Physiol. Sci.*, Montreal, 162–163.

Andersen, P., Dingledine, R., Gjerstad, L., Langmoen, I.A., and Mosfeld-Laursen, A. (1980). Two different responses of hippocampal pyramidal cells to application of gamma-aminobutyric acid. *J. Physiol. 305*, 279–296.

Araki, M., McGeer, P.L., and McGeer, E.G. (1984). Presumptive γ-aminobutyric acid pathways from the midbrain to the superior colliculus studied by a combined horseradish peroxidase-γ-aminobutyric acid transaminase pharmacohistochemical method. *Neuroscience 13*, 433–439.

Armstrong-James, M., and Fox, K. (1987). Spatiotemporal convergence and divergence in the rat S1 "barrel" cortex. *J. Comp. Neurol. 263*, 265–281.

Armstrong-James, M., and George, M.J. (1988). Influence of anesthesia on spontanous activity and receptive field size of single units in rat Sm1 neocortex. *Exp. Neurol. 99,* 369–387.

Asanuma, H. (1975). Recent developments in the study of the columnar arrangement of neurons within the motor cortex *Physiol. Rev. 55,* 143–156.

Asanuma, H., and Rosén, I. (1972). Topographical organization of cortical efferent zones projection to distal forelimb muscles in the monkey. *Exp. Brain Res. 14,* 243–256.

Bauer, R., and Fischer, W.H. (1987). Continuity or incontinuity of orientation columns in visual cortex: A critical evaluation of published and unpublished data. *Neuroscience 22,* 841–847.

Bear, M.F., Schmechel, D.E., and Ebner, F.F. (1985). Glutamic acid decarboxylase in the striate cortex of normal and monoculary deprived kittens. *J. Neurosci. 5,* 1262–1275.

Benjamin, R.M., and Pfaffamann, C. (1955). Cortical localization of taste in albino rat. *J. Neurophysiol. 18,* 59–64.

Bennett, R.E., Ferrington, D.J., and Rowe, M. (1980). Tactile neuron classes within second somatosensory area (SII) of cat cerebral cortex. *J. Neurophysiol. 43,* 292–309.

Berman, N., Payne, B.R., Labar, D.R., and Murphy, E.H. (1982). Functional organization of neurons in the cat striate cortex: Variations in ocular dominance and receptive-field type with cortical laminae and location in the visual field. *J. Neurophysiol. 48,* 1362–1377.

Berman, N.E.J., Wilkes, M.E., and Payne, B.R. (1987). Organization of orientation and direction selectivity in areas 17 and 18 of cat cerebral cortex. *J. Neurophysiol. 58,* 676–699.

Bishop, P.O., Burke, W., and Davis, R. (1962). Single-unit recordings from antidromically activated optic radiation neurons. *J. Physiol. 162,* 432–450.

Bishop, P.O., Coombs, J.S., and Henry, G.H. (1971a). Responses to visual contours: Spatio-temporal aspects of excitation in the receptive fields of simple striate neurones. *J. Physiol. 219,* 625–657.

Bishop, P.O., Coombs, J.S., and Henry, G.H. (1971b). Interaction effects of visual contours on the discharge frequency of simple striate neurones. *J. Physiol. 219,* 659–687.

Bishop, P.O., Coombs, J.S., and Henry, G. H. (1973). Receptive fields of simple cells in the cat striate cortex. *J. Physiol. 231,* 31–60.

Bishop, P.O., Dreher, B., and Henry, G.H. (1972). Simple striate cells: Comparison of responses to stationary and moving stimuli. *J. Physiol. 227,* 15p–17p.

Blakemore, C., and Campbell, F.W. (1969). On the existence of neurones in the human visual system selectively sensitive to the orientation and size of retinal images. *J. Physiol. 203,* 237–260.

Blanton, M.G., Shen, J.M., and Kriegstein, A.R. (1987). Evidence for the inhibitory neurotransmitter γ-aminobutyric acid in aspiny and sparsely spiny nonpyramidal neurons of the turtle dorsal cortex. *J. Comp. Neurol. 259,* 277–297.

Blasdel, G.G., and Fitzpatrick, D. (1984). Physiological organization of layer 4 in macaque striate cortex. *J. Neurosci. 4,* 880–895.

Blasdel, G.G., and Salama, G. (1986). Voltage-sensitive dyes reveal a modular organization in monkey striate cortex. *Nature 321,* 579–585.

Blum, B. (1974). Collateral inhibition in the sensorimotor cortex of the cat and the squirrel monkey—its role in control of neuronal discharges. *Brain Behav. Evol. 9,* 237–247.

Bolz, J., and Gilbert, C.D. (1986). Generation of end-inhibition in the visual cortex via interlaminar connections. *Nature 320,* 362–365.

Brailowski, S., Knight, R. T., Blood, K., and Scabini, D. (1986). Gamma-aminobutyric acid-induced potentiation of cortical hemiplegia. *Brain Res. 362,* 322–330.

Burne, R.A., Parnavelas, J.G., and Lin, C.-S. (1984). Response properties of neurons in the visual cortex of the rat. *Exp. Brain Res. 53,* 374–383.

Burr, D., Morrone, C., and Maffei, L. (1981). Intra-cortical inhibition prevents simple cells from responding to textured visual patterns. *Exp. Brain Res. 43,* 455–458.

Carvell, G.E., and Simons, D.J. (1985). Intracellularly recorded responses of rat SmI cortical neurons to controlled stimulation of the mystacial vibrissae. *Soc. Neurosci. Abst. 11,* 751.

Carvell, G.E., and Simons, D.J. (1987). Thalamocortical response transformation in the vibrissa/barrel system. *Soc. Neurosci. Abst. 13,* 73.8.

Chapin, J.K. (1986). Laminar differences in sizes, shapes, and response profiles of cutaneous receptive fields in the rat SI cortex. *Exp. Brain Res. 62,* 549–559.

Chapin, J.K., and Lin, C.-S. (1984). Mapping the body representation in the S1 cortex of the anaesthetized and awake rat. *J. Comp. Neurol. 229,* 199–213.

Chmielowska, J., Stewart, M.G., and Bourne, R.C. (1988). γ-aminobutyric acid (GABA) immunoreactivity in mouse and rat first somatosensory (SI) cortex: Description and comparison. *Brain Res. 439,* 155–168.

Chow, K.L., Masland, R.H., and Stewart, D.L. (1971). Receptive field characteristics of striate cortical neurons in the rabbit. *Brain Res. 33,* 337–352.

Cleland, B.G., Dubin, M.W., and Levick, E.R. (1971). Sustained and transient neurons in the cat's retina and lateral geniculate nucleus. *J. Physiol. 217,* 473–496.

Cleland, B.G., Levick, W.R., and Wässle, H. (1975). Physiological identification of a morphological class of gangion cells. *J. Physiol. 248,* 151–157.

Connors, B.W., Gutnick, M.J., and Prince, D.A. (1981). Electrophysiological properties of neocortical neurons maintained in vitro. *Soc. Neurosci. Abstr. 7,* 593.

Connors, B.W., Gutnick, M.J., and Prince, D.A. (1982). Electrophysiological properties of neocortical neurons in vitro. *J. Neurophysiol. 48,* 1302–1320.

Connors, B.W., and Kriegstein, A.R. (1986). Cellular physiology of the turtle visual cortex: Distinctive properties of pyramidal and stellate cells. *J. Neurosci. 6,* 164–177.

Constanzo, R.M., and Gardner, E.P. (1980). A quantitative analysis of responses of direction-sensitive neurons in somatosensory cortex of awake monkeys. *J. Neurophysiol. 43,* 1319–1341.

Creutzfeldt, O.D., Kuhnt, U., and Benevento, L.A. (1974). An intracellular analysis of visual cortical neurons to moving stimuli: Responses in a co-operative neuronal network. *Exp. Brain. Res. 21,* 251–274.

Creutzfeldt, O.D., Weber, H., Tanaka, M., and Lee, B.B. (1987). Neuronal representation of spectral and spatial stimulus aspects in foveal and parafoveal area 17 of the awake monkey. *Exp. Brain Res. 68,* 541–564.

Curtis, D.R., and Eccles, R.M. (1958). The excitation of Renshaw cells by pharmacological agents applied electrophoretically. *J. Physiol. 141,* 435–445.

DeFelipe, J., and Jones, E.G. (1985). Vertical organization of gamma-aminobutyric acid-accumulating intrinsic neuronal systems in monkey cerebral cortex. *J. Neurosci. 5,* 3246–3260.

Derrington, A.M., and Lennie, P. (1984). Spatial and temporal contrast sensitivities of neurons in lateral geniculate nucleus of macaque. *J. Physiol. 357,* 219–240.

Deschênes, M., Labelle, A., and Landry, P. (1979). Morphological characterization of slow and fast pyramidal tract cells in the cat. *Brain Res. 178,* 251–274.

DeYoe, E.A., and Van Essen, D.C. (1985). Segregation of efferent connections and receptive field properties in visual area V2 of the macaque. *Nature 317,* 58–61.

DeYoe, E.A., and Van Essen, D.C. (1988). Concurrent processing streams in monkey visual cortex. *Trends Neurobiol.,* in press.

Dichter, M.A. (1980). Physiological identification of GABA as the inhibitory transmitter for mammalian cortical neurons in cell culture. *Brain Res. 190,* 111–121.

Dobbins, A., Zucker, S.W., and Cynader, M.S. (1987). Endstopped neurons in the visual cortex as a substrate for calculating curvature. *Nature 329,* 438–441.

Dräger, U.C. (1975). Receptive fields of single cells and topography in mouse visual cortex. *J. Comp. Neurol. 160,* 269–290.

Dreifuss, J.J., Kelly, J.S., and Krnkević, K. (1969). Cortical inhibition and γ-aminobutyric acid. *Brain Res. 9,* 137–154.

Durham, D., and Woolsey, T.A. (1977). Barrels and columnar cortical organization: Evidence from 2-Deoxyglucose (2-DG) experiments. *Brain Res. 137,* 169–174.

Duysens, J. (1987). Is direction selectivity of cat area 17 cells always independent of contrast and dependent on short-distance interactions? *Exp. Brain Res. 67,* 663–666.

Dykes, R.W. (1983). Parallel processing of somatosensory information: A theory. *Brain Res. Rev. 6,* 47–115.

Dykes, R.W., and Gabor, A. (1981). Magnification factors and organiztion of submodality specific bands in SI cortex of the cat. *J. Comp. Neurol. 202,* 597–620.

Dykes, R.W., and Lamour, Y. (1988). Neurons without demonstrable receptive fields outnumber neurons having receptive fields in samples from the somatosensory cortex of anesthetized or paralyzed cats and rats. *Brain Res. 440,* 133–143.

Dykes, R.W., Lamour, Y., Diadori, P., Landry, P., and Dutar, P. (1988). Somatosensory cortical neurons with identifiable electrophysiological signature. *Brain Res. 441,* 48–58.

Dykes, R.W., Landry, P., Metherate, R., and Hicks, T.P. (1984). Functional role of GABA in cat primary somatosensory cortex: Shaping receptive fields of cortical neurons. *J. Neurophysiol. 52,* 1066–1093.

Dykes, R.W., Rasmusson, D.D., and Hoeltzell, P. (1980). Organization of primary somatosensory cortex in the cat. *J. Neurophysiol. 43,* 1527–1546.

Emerson, R.C., Citron, M.C., Vaughn, W.J., and Klein, S.A. (1987). Non-linear directionally selective subunits in complex cells of cat striate cortex. *J. Neurophysiol. 58,* 33–65.

Enroth-Cugell, C., and Robson, J.G. (1966). The contrast sensitivity of retinal ganglion cells of the cat. *J. Physiol. 187,* 517–552.

Esclapez, M., Campistron, G., and Trottier, S. (1987). Immunucytochemical localization and morphology of GABA-containing neurons in the prefrontal and frontoparietal cortex of the rat. *Neurosci. Lett. 77,* 131–136.

Eysel, U.T., Wörgötter, F., and Pape, H.-C. (1987). Local cortical lesions abolish lateral inhibition at direction selective cells in cat visual cortex. *Exp. Brain Res. 68,* 606–612.

Favorov, O.V., Diamond, M.E., and Whitsel, B.L. (1987). Evidence for a mosaic representation of the body surface in area 3b of the somatic cortex of cat. *Proc. Natl. Acad. Sci. USA 84,* 6606–6610.

Favorov, O.V., and Whitsel, B.L. (1988). Spatial organization of the peripheral input to area 1 cell columns. II. The forelimb representation achieved by a mosaic of segregates. *Brain Res. Rev. 13,* 43–56.

Ferrington, D.G., and Rowe, M. (1980). Differential contributions to coding of cutaneous vibratory information by cortical somatosensory areas I and II. *J. Neurophysiol. 43,* 310–331.

Ferster, D., and Koch, C. (1987). Neuronal connections underlying orientation selectivity in cat visual cortex. *T.I.N.S. 10,* 487–493.

Ferster, D., and LeVay, S. (1978). The axonal arborizations of lateral geniculate neurons in the striate cortex of the cat. *J. Comp. Neurol. 182,* 923–944.

Ferster, D., and Lindström, S. (1983). An intracellular analysis of geniculo-cortical connectivity in area 17 of the cat. *J. Physiol. 342,* 181–215.

Ferster, D., and Lindström, S. (1985). Synaptic excitation of neurones in area 17 of the cat by intracortical axon collaterals of cortico-geniculate cells. *J. Physiol. 367,* 233–252.

Fitzpatrick, D., Lund, J.S., Schmechel, D.E., and Towles, A.C. (1987). Distribution of GABAergic neurons and axon terminals in the macaque striate cortex. *J. Comp. Neurol. 264,* 73–91.

Freund, T.F., Martin, K.A.C., Smith, A.D., and Somogyi, P. (1983). Glutamate decarboxylase-immunoreactive terminals of Golgi-impregnated axoaxonic cells and of presumed basket cells in synaptic contact with pyramidal neurons of the cat's visual cortex. *J. Comp. Neurol. 221,* 263–278.

Freund, T.F., Martin, K.A.C., Somogyi, P., and Whitteridge, D. (1985b). Innervation of cat visual areas 17 and 18 by physiologically identified X- and Y-type afferents. II. Identification of postsynaptic targets by GABA immunocytochemistry and Golgi impregnation. *J. Comp. Neurol. 242,* 275–291.

Freund, T.F., Martin, K.A.C. and Whitteridge, D. (1985a). Innervation of cat visual areas 17 and 18 by physiologicallyidentified X- and Y-type afferents. I. Arborization patterns and quantitative distribution of postsynaptic elements. *J. Comp. Neurol. 242,* 263–274.

Fukuda, Y., and Stone, J. (1975). Direct identification of the cell bodies of Y-, X-, and W-cells in the cat's retina. *Vision Res. 15,* 1034–1036.

Gabbott, P.L.A., and Somogyi, P. (1986). Quantitative distribution of GABA-immunoreactive neurons in the visual cortex (area 17) of the cat. *Exp. Brain Res. 61,* 323–331.

Gardner, E.P., and Costanzo, R.M. (1980a). Temporal integration of multiple-point stimuli in primary somatosensory cortical receptive fields of alert monkeys. *J. Neurophysiol. 43,* 444–468.

Gardner, E.P., and Costanzo, R.M. (1980b). Neuronal mechanisms underlying direction sensitivity of somatosensory cortical neurons in awake monkeys. *J. Neurophysiol. 43,* 1342–1354.

Giaretta, D., Kostopoulos, G., Gloor, P., and Avoli, M. (1985). Intracortical inhibitory machanisms are preserved in feline generalized penicillin epilepsy. *Neurosci. Lett. 59,* 203–208.

Gilbert, C.D. (1977). Laminar differences in receptive field properties of cells in cat primary visual cortex. *J. Physiol. 268,* 391–421.

Gilbert, C.D., and Wiesel, T.N. (1979). Morphology and intracortical projections of functionally characterised neurones in the cat visual cortex. *Nature 280,* 120–125.

Gilbert, C.D., and Wiesel, T.N. (1983). Laminar specialization and intracortical connections in cat primary visual cortex. In: *The Organization of the Cerebral Cortex* (G. Adelman, S.G. Dennis, F.O. Schmidt, F.G. Worden, eds.). MIT Press, Cambridge, Mass., pp. 163–191.

Goodwin, A.W., and Henry, G.H. (1975b). Direction selectivity of complex cells in a comparison with simple cells. *J. Neurophysiol. 38,* 1524–1540.

Goodwin, A.W., Henry, G.H., and Bishop, P.O. (1975a). Direction selectivity of simple striate cells: Properties and mechanism. *J. Neurophysiol. 38,* 1500–1523.

Gulyas, B., Orban, G.A., Duysens, J., and Maes, H. (1987). The suppressive influence of moving textured backgrounds on responses of cat striate neurons to moving bars. *J. Neurophysiol. 57,* 1767–1791.

Gutnick, M.J., Connors, B.W., and Ransom, B.R. (1981). Dye coupling between glial cells in the guinea pig neocortical slice. *Brain Res. 213,* 486–492.

Haberly, L.B., Hansen, D.J., Feig, S.L., and Presto, S. (1987). Distribution and ultrastructure of neurons in opossum piriform cortex displaying immunoreactivity to GABA and GAD and high-affinity tritiated GABA uptake. *J. Comp. Neurol. 266,* 269–290.

Haenny, P.E., and Schiller, P.H. (1988). State dependent activity in monkey visual cortex. I. Single cell activity in V1 and V4 on visual tasks. *Exp. Brain Res. 69,* 225–244.

Hamos, J.E., Davis, T.L., and Sterling, P. (1983). Four types of neurons in layer IVab of cat cortical area 17 accumulate ^{3}H-GABA. *J. Comp. Neurol. 217,* 449–457.

Hand, P.J. (1981). The 2-Deoxyglucose method. In: *Neuroanatomical Tract-Tracing Methods* (L. Heimer and M. J. Robards, eds.). Plenum Press, New York, pp. 511–538.

Harvey, A.R. (1980). A physiological analysis of subcortical and commissural projections of areas 17 and 18 of the cat. *J. Physiol. 302,* 507–534.

Hellweg, F.-C., Schutz, W., and Creutzfeldt, O.D. (1977). Extracellular and intracellular recordings from cat's cortical whisker projection area: Thalamocortical response transformation. *J. Neurophysiol. 40,* 463–479.

Hendry, S.H.C., Schwark, H.D., Jones, E.G., and Yan, J. (1987). Numbers and proportions of GABA immunoreactive neurons in different areas of monkey cerebral cortex. *J. Neurosci. 7,* 1503–1519.

Henry, G.H. (1977). Receptive field classes of cells in the striate cortex of the cat. *Brain Res. 133,* 1–28.

Henry, G.H. (1985). Physiology of cat striate cortex. In: *The Cerebral Cortex.* Vol. 3, Visual Cortex (A. Peters and E.G. Jones, eds.). Plenum Press, New York, pp. 119–155.

Henry, G.H., Harvey, A.R., and Lund, J.S. (1979). The afferent connections and laminar distribution of cells in the cat striate cortex. *J. Comp. Neurol. 187,* 725–744.

Heywood, C.A., and Cowey, A. (1987). On the role of cortical are V4 in the discrimination of hue and pattern in macaque monkeys. *J. Neurosci. 7,* 2601–2617.

Hicks, T.P., and Dykes, R.W. (1983). Receptive field size of certain neurons in primary somatosensory cortex is determined by GABA-mediated intracortical inhibition. *Brain Res. 274,* 160–164.

Hicks, T.P., Landry, P., Metherate, R., and Dykes, R.W. (1985). Functional properties of neurons mediated by GABA in cat somatosensory cortex under barbiturate and urethane anesthesia. In: *Development, Organization, and Processing in Somatosensory Pathways* (M. Rowe and W.D. Willis, eds.). Alan R. Liss, New York, 265–276.

Hillman, H.H., Campbell, W.J., and McIlwain, H. (1963). Membrane potential in isolated and electrically stimulated mammalian cerebral cortex: Effects of chlorpromazine, cocaine, phenobarbitone and protamine on the tissues' electrical and chemical responses to stimulation. *J. Neurochem. 10,* 325–339.

Hoffmann, K.P., and Stone, J. (1971). Conduction velocity of afferents to cat visual cortex: A correlation with cortical receptive field properties. *Brain Res.* 32, 460–466.

Hoffmann, K.P., Stone, J., and Sherman, S.M. (1972). Relay of receptive field properties in dorsal lateral geniculate nucleus of the cat. *J. Neurophysiol. 35,* 518–531.

Horton, J.C., and Hubel, D.H. (1981). A regular patchy distribution of cytochrome oxidase staining in primary visual cortex of the macaque monkey. *Nature 292,* 762–764.

Houser, C.R., Hendry, S.H.C., Jones, E.G., and Vaughn, J.E. (1983). Morphological diversity of immunocytochemically identified GABA neurons in the monkey sensory-motor cortex. *J. Neurocytol. 12,* 617–638.

Houser, C.R., Vaughn, J.E., Hendry, S.H.C., Jones, E.G., and Peters, A. (1984). GABA neurons in the cerebral cortex. In: *The Cerebral Cortex.* Vol. 2, Functional Properties of Cortical Cells (E.G. Jones and A. Peters, eds.). Plenum Press, New York, pp. 63–89.

Hubel, D.H., and Livingstone, M.S. (1987). Segregation of form, color, and stereopsis in primate area 18. *J. Neurosci. 7,* 3378–3415.

Hubel, D.H., and Wiesel, T.N. (1961). Integrative action in the cat's lateral geniculate body. *J. Physiol. 155,* 385–398.

Hubel, D.H., and Wiesel, T.N. (1962). Receptive fields, binocular interaction and functional architecture in the cat's visual cortex. *J. Physiol. 160,* 106–154.

Hubel, D.H., and Wiesel, T.N. (1965). Receptive fields and functional architecture in two non-striate visual areas (18 and 19) of the cat. *J. Neurophysiol. 28,* 229–289.

Hubel, D.H., and Wiesel, T.N. (1968). Receptive fields and functional architecture of monkey striate cortex. *J. Physiol. 195,* 215–243.

Hubel, D.H., and Wiesel, T.N. (1969). Anatomical demonstration of columns in monkey striate cortex. *Nature 221,* 747–750.

Hubel, D.H., and Wiesel, T.N. (1972). Laminar and columnar distribution of geniculo-cortical fibers in the Macaque monkey. *J. Comp. Neurol. 146,* 421–450.

Hubel, D.H., and Wiesel, T.N. (1977). Functional architecture of macaque monkey visual cortex. *Proc. R. Soc.* Ser. B. 198, 1–59.

Hubel, D., Wiesel, T.N., and Stryker, P.M. (1978). Anatomical demonstration of orientation columns in macaque monkey. *J. Comp. Neurol. 177,* 361–380.

Humphrey, A.L., Sur, M., Uhlrich, D.J., and Sherman, S.M. (1985). Projection patterns of individual X- and Y-cell axons from the lateral geniculate nucleus to cortical area 17 in the cat. *J. Comp. Neurol. 233,* 159–189.

Humphrey, A.L., and Weller, R.E. (1988a). Functionally distinct groups of X-cells in the lateral geniculate nucleus of the cat. *J. Comp. Neurol. 268,* 429–447.

Humphrey, A.L., and Weller, R.E. (1988b). Structural correlates of functionally distinct X-cells in the lateral geniculate nucleus of the cat. *J. Comp. Neurol. 268,* 448–468.

Hyvärinen, J., Poranen, A., and Jokinen, Y. (1980). Influence of attentive behavior on neuronal responses to vibration in primary somatosensory cortex of the monkey. *J. Neurophysiol. 43,* 870–882.

Imig, T.J., and Adrián, H.O. (1977). Binaural columns in the primary field (AI) of cat auditory cortex. *Brain Res. 138,* 241–257.

Ito, M. (1985). Processing of vibrissa sensory information within the rat neocortex. *J. Neurophysiol. 54,* 479–490.

Iversen, L.L., Mitchell, J.F., and Srinivansan, V. (1971). The release of gamma-aminobutyric acid during inhibition in the cat visual cortex. *J. Physiol. 212,* 519–534.

Iwamura, Y., and Tanaka, M. (1978). Functional organization of receptive fields in the cat somatosensory cortex. I. Integration within the coronal region. *Brain Res. 151,* 49–60.

Jennings, V. (1979). Function and morphology of neurons in the vibrissal region of rat somatosensory neocortex. Ph.D. dissertation, University of Rochester.

Juliano, S.L., and Whitsel, B.L. (1987). A combined 2-deoxyglucose and neurophysiological study of primate somatosensory cortex. *J. Comp. Neurol. 263,* 514–525.

Kang, R., Herman, D., MacGillis, M., and Zarzecki, P. (1985). Convergence of sensory inputs in somatosensory cortex: interactions from seperate afferent sources. *Exp. Brain Res. 57,* 271–278.

Kato, H., Bishop, P.O., and Orban, G.A. (1978). Hypercomplex and simple/complex cell classifications in cat striate cortex. *J. Neurophysiol. 41,* 1071–1095.

Kaufman, P.E., Wallingford, R., Ostdahl, R., and Somjen, G. (1973). Receptive field properties of visual cortical neurons in the tree shrew *(Tupaia glis). Soc. Neurosci. Abst. 3,* 179.

Keller, A., and White, E.L. (1986). Distribution of glutamic acid decarboxylase-immunoreactive structures in the barrel region of mouse somatosensory cortex. *Neurosci. Lett. 66,* 245–250.

Keller, A., and White, E.L. (1987). Synaptic organization of GABAergic neurons in the mouse SmI cortex. *J. Comp. Neurol. 262,* 1–12.

Keller, A., and White, E.L. (1989). Triads: A synaptic network component in the cerebral cortex. *Brain Res.,* in press.

Kelly, J.P., and Van Essen, D.C. (1974). Cell structure and function in the visual cortex of the cat. *J. Physiol. 238,* 515–547.

Kelly, J.S., and Renaud, L.P. (1974). Physiological identification of inhibitory interneurons in the feline pericruciate cortex. *Neuropharmacology 13,* 463–474.

Kemp, J.A. (1983). Intracellular recordings from rat visual cortical cells in vitro and the action of GABA. *J. Physiol. 349,* 13P.

Kisvarday, Z.F., Martin, K.A.C., Whitteridge, D., and Somogyi, P. (1985). Synaptic connections of intracellularly filled clutch cells: A type of small basket cell in the visual cortex of the cat. *J. Comp. Neurol. 241,* 111–137.

Kisvarday, Z.F., Cowey, A., and Somogyi, P. (1986). Synaptic relationships of a type of GABA-immunoreactive neuron (clutch cell), spiny stellate cells and lateral geniculate nucleus afferents in layer IVC of the monkey striate cortex. *Neuroscience 19,* 741–761.

Kisvarday, Z.F., Martin, K.A.C., Friedlander, M.J., and Somogyi, P. (1987). Evidence for interlaminar inhibitory circuits in the striate cortex of the cat. *J. Comp. Neurol. 260,* 1–19.

Koch, C., Poggio, T., and Torre, V. (1982). Retinal ganglion cells: A functional interpretation of dendritic morphology. *Phil. Trans. R. Soc.* Ser. B 298, 227–264.

Kriegstein, A.R. (1987). Synaptic responses of cortical pyramidal neurons to light stimulation in the isolated turtle visual system. *J. Neurosci. 7,* 2488–2492.

Kriegstein, A.R., and Connors, B.W. (1986). Cellular physiology of the turtle visual cortex: Synaptic properties and intrinsic circuitry. *J. Neurosci. 6,* 178–191.

Krnjević, K. (1984). Neurotransmitters in cerebral cortex: A general account. In: *The Cerebral Cortex.* Vol. 2, Functional Properties of Cortical Cells (E.G. Jones and A. Peters, eds.). Plenum Press, New York, pp. 39–61.

Krnjević, K., and Schwartz, S. (1967). The action of γ-aminobutyric acid on cortical neurones. *Exp. Brain Res. 3,* 320–336.

Kuffler, S.W. (1953). Discharge patterns and functional organization of mammalian retina. *J. Neurophysiol. 16,* 37–68.

Lee, B.B., Cleland, B.G., and Creutzfeldt, O.D. (1977). The retinal input to cells in area 17 of the cat's cortex. *Exp. Brain Res. 30,* 527–538.

Lennie, P. (1980). Parallel visual pathways: A review. *Vision Res. 20,* 561–594.

LeVay, S., and Ferster, D. (1977). Relay cell classes in the lateral geniculate nucleus of the cat and the effects of visual deprivation. *J. Comp. Neurol. 172,* 563–584.

LeVay, S., Hubel, D.H., and Wiesel, T.N. (1975). The pattern of ocular dominance columns in macaque visual cortex revealed by a reduced silver stain. *J. Comp. Neurol. 159,* 559–576.

LeVay, S., McConnell, S.K., and Luskin, M.B. (1987). Functional organization of primary visual cortex in the mink (mustela vison), and a comparison with the cat. *J. Comp. Neurol. 257,* 422–441.

LeVay, S., Stryker, M., and Shatz, C. (1978). Ocular dominance columns and their development in layer IV of the cat's visual cortex: A quantitative study. *J. Comp. Neurol. 178,* 223–244.

Leventhal, A.G., Rodieck, R.W., and Dreher, B. (1981). Retinal ganglion cell classes in the old world monkey: Morphology and central projections. *Science 213,* 1139–1142.

Lin, C.-S., Friedlander, M.J., and Sherman, S.M. (1979). Morphology of physiologically identified neurons in the visual cortex of the cat. *Brain Res. 172,* 344–348.

Lin, C.-S., Lu, S.M., and Schmechel, D.E. (1985). Glutamic acid decarboxylase immunoreactivity in layer IV of barrel cortex of rat and mouse. *J. Neurosci. 5,* 1934–1939.

Livingstone, M.S., and Hubel, D.H. (1984a). Specifity of intrinsic connections in primate primary visual cortex. *J. Neurosci. 4,* 2830–2835.

Livingstone, M.S., and Hubel, D.H. (1984b). Anatomy and physiology of a color system in the primate visual cortex. *J. Neurosci. 4,* 309–356.

Livingstone, M.S., and Hubel, D.H. (1987a). Connections between layer 4B of area 17 and the thick cytochrome oxidase stripes of area 18 in the squirrel monkey. *J. Neurosci. 7,* 3371–3377.

Livingstone, M.S., and Hubel, D.H. (1987b). Psychophysical evidence for separate channels for the perception of form, color, movement, and depth. *J. Neurosci. 7,* 3416–3468.

Lorente de Nó, R. (1938). Cerebral Cortex: Architecture, intracortical connections, motor projections. In: *The Physiology of the Nervous System,* 3rd. ed (J.F. Fulton, ed.). Oxford University Press, London, pp. 291–325.

Löwel, S., Freeman, B., and Singer, W. (1987). Topographic organization of the orientation column system in large flat-mounts of the cat visual cortex: A 2-deoxyglucose study. *J. Comp. Neurol.* *255*, 401–415.

Lund, J.S. (1988). Anatomical organization of macaque monkey striate visual cortex. *Ann. Rev. Neurosci.* *11*, 253–288.

Lund, J.S., Lund, R.D., Hendrickson, A.E., Bunt, A.H., and Fuchs, A.F. (1975). The origin of efferent pathways from the primary visual cortex, area 17, of the Macaque monkey as shown by retrograde transport of horesradish peroxidase. *J. Comp. Neurol.* *164*, 287–304.

Maffei, L. (1985). Complex cells control simple cells. In: *Models of the Visual Cortex* (D. Rose and V.G. Dobson, eds.). Wiley, New York, pp. 334–340.

Malpeli, J.G. (1983). Activity of cells in area 17 of the cat in absence of input from layer A of lateral geniculate nucleus. *J. Neurophysiol.* *49*, 595–610.

Malpeli, J.G., Lee, C., Schwark, H.D., and Weyand, T.G. (1986). Cat area 17. I. Pattern of thalamic control of cortical layers. *J. Neurophysiol.* *56*, 1062–1073.

Malpeli, J.G., Schiller, P.H., and Colby, C.L. (1981). Response peoperties of single cells in monkey striate cortex during reversible inactivaton of individual lateral genuculate laminae. *J. Neurophysiol.* *46*, 1102–1119.

Mangini, N.J., and Pearlman, A.L. (1980). Laminar distribution of receptive field properties in the primary visual cortex of the mouse. *J. Comp. Neurol.* *193*, 203–222.

Marshall, W.H., and Talbot, S.A. (1942). Recent evidence for neural mechanisms in vision leading to a general theory of sensory acuity. In: *Visual Mechanisms* (H. Kluver, ed.). Cattell, Lancaster, Pa., pp. 117–164.

Martin, J.H., and Spencer, A. (1985). Mechanoreceptive submodality channel interactions: Single unit analysis of afferent inhibition in the primary somatosensory cortex of the cat. *Brain Res.* *327*, 279–288.

Martin, K.A.C. (1984). Neuronal Circuits in cat striate cortex. In: *The Cerebral Cortex.* Vol. 2, Functional Properties of Cortical Cells (E.G. Jones and A. Peters, eds.). Plenum Press, New York, pp. 241–284.

Martin, K.A.C., Somogyi, P., and Whitteridge, D. (1983). Physiological and morphological properties of identified basket cells in the cat's visual cortex. *Exp. Brain Res.* *50*, 193–200.

Martin, K.A.C., and Whitteridge, D. (1984a). Form, function and intracortical projections of spiny neurones in the striate visual cortex of the cat. *J. Physiol.* *353*, 463–504.

Martin, K.A.C., and Whitteridge, D. (1984b). The relationship of receptive field properties to the dendritic shape of neurones in the cat striate cortex. *J. Physiol.* *356*, 291–302.

Mastronarde, D.M. (1987a). Two classes of single-input X-cells in the cat lateral geniculate nucleus. I. Receptive-field properties and classification of cells. *J. Neurophysiol.* *57*, 357–380.

Mastronarde, D.M. (1987b). Two classes of single-input X-cells in the cat lateral geniculate nucleus. II. Retinal inputs and the generation of receptive-field properties. *J. Neurophysiol.* *57*, 381–413.

Matsubara, J.A., Cynader, M.S., and Swindale, N.V. (1987). Anatomical properties and physiological correlates of the intrinsic connections in cat area 18. *J. Neurosci.* *7*, 1428–1446.

Matsubara, J.A., Cynader, M.S., Swindale, N.V., and Stryker, M.P. (1985). Intrinsic projections within visual cortex: Evidence for orientation specific local connections. *Proc. Natl. Acad. Sci. USA 82,* 935–939.

Matsubara, J.A., and Phillips, D.P. (1988). Intracortical connections and their physiological correlates in the primary auditory cortex (AI) of the cat. *J. Comp. Neurol. 268,* 38–48.

McCormick, D.A., Connors, B.W., Lighthall, J.W., and Prince, D.A. (1985). Comparative electrophysiology of pyramidal and sparsely spiny stellate neurons of the neocortex. *J. Neurophysiol. 54,* 782–805.

McCourt, M.E., Boyapati, J., and Henry, G.H. (1987). Properties of area 17/18 border neurons contributing to the visual transcallosal pathway in the cat. *Soc. Neurosci. Abst. 13,* 178.18.

Meinecke, D.L., and Peters, A. (1987). GABA immunoreactive neurons in rat visual cortex. *J. Comp. Neurol. 261,* 388–404.

Metherate, R., and Dykes, R.W. (1985). Simultanous recordings from pairs of cat somatosensory cortical neurons with overlapping peripheral receptive fields. *Brain Res. 341,* 119–129.

Michael, C.R. (1978). Color vision mechanisms in monkey striate cortex: Simple cells with dual opponent-color receptive fields. *J. Neurophysiol. 41,* 1233–1249.

Michael, C.R. (1987). Comparative studies of the color cells in layer IVCß and in the blobs of the monkey's striate cortex. *Soc. Neurosci. Abst. 13,* 5.42.

Michalski, A., Gerstein, G.L., Czarkowska, J., and Tarnecki, R. (1983). Interactions between cat striate cortex neurons. *Exp. Brain Res. 51,* 97–107.

Miller, J.M., Sutton, D., Pfingst, B., Ryan, A., and Beaton, R. (1972). Single cell activity in the auditory cortex of rhesus monkeys: Behavioral dependency. *Science 177,* 449–451.

Misgeld, U., Deisz, R.A., Dodt, H.U., and Lux, H.D. (1986). The role of chloride transport in postsynaptic inhibition of hippocampal neurons. *Science 232,* 1413–1415.

Mitani, A., Minoru, S., Kazuo, I., Sakashk, N., Motoi, K., and Mizuno, N. (1985). Morphology and laminar organization of electrophysiologically identified neurons in the primary auditory cortex in the cat. *J. Comp. Neurol. 235,* 430–477.

Morrone, M.C., Burr, D.C., and Speed, H.D. (1987). Cross-orientation inhibition in cat is GABA mediated. *Exp. Brain Res. 67,* 635–644.

Mountcastle, V.B. (1957). Modality and topographic properties of single neurons of cat's somatic sensory cortex. *J. Neurophysiol. 20,* 408–434

Mountcastle, V.B. (1979). An organizing principle for cerebral function: The unit module and the distributed system. In: *The Neurosciences, Fourth Study Program* (F.O. Schmidt and F.G. Worden, eds.). MIT Press, Cambridge, Mass., pp. 21–42.

Mountcastle, V.B., Motter, B.C., Steinmetz, M.A., and Sestokas, A.K. (1987). Common and differential effects of attentive fixation on the excitability of parietal and prestriate (V4) cortical visual neurons in the macaque monkey. *J. Neurosci. 7,* 2239–2255.

Mountcastle, V.B., and Powell, T.P.S. (1959). Neural mechanisms subserving cutaneous sensibility, with special reference to the role of afferent inhibition in sensory perception and discrimination. *Bull. John Hopkins Hosp. 105,* 201–232.

Mountcastle, V.B., Talbot, W.H., Sakata, H., and Hyvärinen, J. (1969). Cortical neuronal mechanisms in flutter-vibration studied in unanesthetized monkeys. Neuronal periodicity and frequency discrimination. *J. Neurophysiol. 32,* 452–484.

Movshon, J.A. (1974). Velocity preferences of simple and comples cells in the cat's striate cortex. *J. Physiol. 242,* 121P–123P.

Mugnaini, E., and Oertel, W.H. (1985). An atlas of the distribution of GABAergic neurons and terminals in the rat CNS as revealed by GAD immunohistochemistry. In: *Handbook of Chemical Neuroanatomy.* Vol. 4, GABA and Neuropeptides in the CNS, Part I (A. Bjorklund and T. Hokfelt, eds.). Elsevier Science Publishers B.V., Amsterdam, pp. 436–608.

Nagai, T., McGeer, P.L., and McGeer, E.G. (1983). Distribution of GABA-T-intensive neurons in the rat forebrain and midbrain. *J. Comp. Neurol. 218,* 220–238.

Nelson, J.I., and Frost, B.J. (1978). Orientation-selective inhibition from beyond the classic visual receptive field. *Brain Res. 139,* 359–365.

Noda, T., and Yamamoto, T. (1984). Response properties and morphological identification of neurons in the cat motor cortex. *Brain Res. 306,* 197–206.

Norton, T.T., Rager, G., and Kretz, R. (1985). On and OFF regions in layer IV of striate cortex. *Brain Res. 327,* 319–323.

Orban, G.A., Gulyas, B., and Vogels, R. (1987). Influence of a moving textured background on direction selectivity of cat striate neurons. *J. Neurophysiol. 57,* 1792–1812.

Ottersen O.P., and Storm-Mathisen, J. (1984a). Glutamate and GABA-containing neurons in the mouse and rat brain, as demonstrated with a new immunocytochemical technique. *J. Comp. Neurol. 229,* 374–392

Ottersen O.P., and Storm-Mathisen, J. (1984b). Immunocytochemical visualization of GABA fixed by glutaraldehyde in brain tissue. *Neuropharmacology 23,* 855–857.

Palmer, L.A., and Rosenquist, A.C. (1974). Visual receptive fields of single striate cortical units projecting to the superior colliculus in the cat. *Brain Res. 67,* 27–42.

Parnavelas, J.G. (1984). Physiological properties of identified neurons. In: *The Cerebral Cortex.* Vol. 2, Cellular Components of the Cerebral Cortex (E.G. Jones and A. Peters, eds.). Plenum Press, New York, pp. 205–239.

Parnavelas, J.G., Burne, R.A., and Lin, C.-S. (1981). Receptive field properties of neurons in the visual cortex of the rat. *Neurosci. Lett. 27,* 291–296.

Parnavelas, J.G., Burne, R.A., and Lin C.-S. (1983). Distribution and morphology of functionally identified neurons in the visual cortex of the rat. *Brain Res. 261,* 21–29.

Penfield, W., and Boldrey, E. (1937). Somatic motor and sensory representation in the cerebral cortex of man as studied by electrical stimulation. *Brain 60,* 389–443.

Penny, J.B., and Young, A.B. (1982). Quantitative autoradiography of neurotransmitter receptors in Huntington's disease. *Neurology 32,* 1391–1395.

Perkel, D.H., Gerstein, G.L., and Moore, G.P. (1967). Neuronal spike trains and stochastic point processes. I. The single spike train. *Biophys. J. 7,* 391–418.

Peters, A., Kara, D.A., and Harriman, K.M. (1985). The neuronal composition of area 17 of rat visual cortex. III. Numerical considerations. *J. Comp. Neurol. 238,* 263–274.

Pettigrew, J.D., and Daniels, J.D. (1973). Gamma-aminobutyric acid antagonism in visual cortex: Different effects on simple, complex, and hypercomplex neurons. *Science 5,* 81–83.

Pettigrew, J.D., Nikara, T., and Bishop, P.O. (1968). Responses to moving slits by single units in cat striate cortex. *Exp. Brain Res. 6,* 373–390.

Powell, T.P.S., and Mountcastle, V.B. (1959). The cytoarchitecture of the postcentral gyrus of the monkey macaca mulatta. *Bull. Johns Hopkins Hosp. 105,* 108–132.

Ramoa, A.S., Shadlen, M., Skottun, B.C., and Freeman, R.D. (1986). A comparison of inhibition in orientation and spatial frequency selectivity of cat visual cortex. *Nature 321,* 237–239.

Ribak, C.E. (1978). Aspinous and sparsely-spinous stellate neurons in the visual cortex of rats contain glutamic acid decarboxylase. *J. Neurocytol. 7,* 461–478.

Ribak, C.E., Harris, A.B., Vaughn, J.E., and Roberts, E. (1979). Inhibitory, GABAergic nerve terminals decrease at sites of focal epilepsy. *Science 205,* 211–214.

Ribak, C.E., Hunt, C.A., Bakay, R.A.E., and Oertel, W.H. (1986). A decrease in the nuber of GABAergic somata is associated with the preferential loss of GABAergic terminals at epileptic foci. *Brain Res. 363,* 78–90.

Rocha-Miranda, C.E., Bombardieri, R.A., de Monasterio, F.M., and Linden, R. (1973). Receptive fields in the visual cortex of the opposum. *Brain Res. 63,* 362–367.

Rose, D., and Blackmore, C. (1974). Effects of bicuculline on functions of inhibition in visual cortex. *Nature 249,* 375–379.

Rosén, I., and Asanuma, H. (1972). Peripheral afferent inputs to the forelimb area of the monkey motor cortex: Input output relations. *Exp. Brain Res. 14,* 257–273.

Schall, J.D., Vitek, D.J., and Leventhal, A.G. (1986). Retinal constraints on orientation specificity in cat visual cortex. *J. Neurosci. 6,* 823–836.

Scharfman, H.E., and Sarvey, J.M. (1985). Responses to gamma-aminobutyric acid applied to call bodies and dendrites of rat visual cortical neurons. *Brain Res. 358,* 385–389.

Scharfman, H.E., and Sarvey, J.M. (1987). Responses to GABA recorded from identified rat visual cortical neurons. *Neurosci. 23,* 407–422.

Schiller, P.H., Finlay, B.L., and Volman, S.F. (1976a). Quantitative studies of single-cell properties in monkey striate cortex. I. Spatiotemporal organization of receptive fields. *J. Neurophysiol. 39,* 1288–1319.

Schiller, P.H., Finlay, B.L., and Volman, S.F. (1976b). Quantitative studies of single-cell properties in monkey striate cortex. II. Orientation specificity and ocular dominance. *J. Neurophysiol. 39,* 1320–1333.

Schiller, P.H., Finlay, B.L., and Volman, S.F. (1976c). Quantitative studies of single-cell properties in monkey striate cortex. III. Spatial frequency. *J. Neurophysiol. 39,* 1334–1351.

Schwark, H.D., Malpeli, J.G., Weyand, T.G., and Lee, C. (1986). Cat area 17. II. Response properties of infragranular layer neurons in the absence of supragranular layer activity. *J. Neurophysiol.* *56,* 1074–1087.

Segev, I., and Parnas, I. (1983). Synaptic integration mechanisms. Theoretical and experimental investigation of temporal and postsynaptic interactions between excitatory and inhibitory inputs. *Biophys. J.* *41,* 41–50.

Shatz, C.J., Lindström, S., and Wiesel, T.N. (1977). The distribution of afferents representing the right and left eyes in the cat's visual cortex. *Brain Res.* *131,* 103–116.

Sherman, S.M., Watkins, D.W., and Wilson, J.R. (1976). Further differences in receptive field properties of simple and complex cells in cat striate cortex. *Vision Res.* *16,* 919–927.

Shipp, S., and Zeki, S. (1985). Segregation of pathways leading from area V2 to areas V4 and V5 of macaque monkey visual cortex. *Nature* *315,* 322–325.

Sillito, A.M. (1975). The contribution of inhibitory mechanisms to the receptive field properties of neurones in the striate cortex of the cat. *J. Physiol.* *250,* 305–329.

Sillito, A.M. (1977). Inhibitory processes underlying the directional specificity of simple, complex and hypercomplex cells in the cat's visual cortex. *J. Physiol.* *271,* 699–720.

Sillito, A.M. (1984). Functional considerations of the operation of GABAergic inhibitory processes in the visual cortex. In: *Cerebral Cortex.* Vol. 2, Functional Properties of Cortical Cells (E. G. Jones and A. Peters, eds.). Plenum Press, New York, pp. 91–117.

Sillito, A.M., Kemp, J.A., and Patel, H. (1980). Inhibitory interactions contributing to the ocular dominance of monocularly dominated cells in the normal cat visual cortex. *Exp. Brain Res.* *41,* 1–10.

Simmons, P.A., and Pearlman, A.L. (1983). Receptive-field properties of transcallosal visual cortical neurons in the normal and reeler mouse. *J. Neurophysiol.* *50,* 838–848.

Simons, D.J. (1978). Response properties of vibrissa units in rat SI somatosensory neocortex. *J. Neurophysiol.* *41,* 798–820.

Simons, D.J. (1983). Multi-whisker stimulation and its effects on vibrissa units in rat SmI barrel cortex. *Brain Res.* *276,* 178–182.

Simons, D.J. (1985). Temporal and spatial integration in the rat SI vibrissa cortex. *J. Neurophysiol.* *54,* 615–635.

Simons, D.G., and Carvell, G.E. (1989). Thalamocortical response transformation in rat vibrissa/barrel system. *J. Neurophysiol.,* *61,* 311–330.

Simons, D.J., Carvell, G.E., and Land, P.W. (1989). The vibrissa/barrel cortex as a model of sensory information precessing. In: *Sensory Processing in Mammalian Brain: Neuronal Substrate and Experimental Strategy* (J.S. Lund, ed.). Oxford Univ. Press, New York, pp. 67–83.

Simons, D.J., and Woolsey, T.A. (1979). Functional organization in mouse barrel cortex. *Brain Res.* *165,* 327–332.

Singer, W., Tretter, F., and Cynader, M. (1975). Organization of cat striate cortex: A correlation of receptive-field properties with afferent and efferent connections. *J. Neurophysiol.* *38,* 1080–1098.

Skeen, L.C., Humphrey, A.L., Norton, T.T., and Hall, W.C. (1978). Deoxyglucose mapping of the orientation column system in the striate cortex of the tree shrew, *Tupaia glis. Brain Res. 142,* 538–545.

Solberg, Y., White, E.L., and Keller, A. (1988). Types and distribution of glutamic acid decarboxylase (GAD)-immunoreactive neurons in mouse motor cortex. *Brain Res. 459,* 168–172.

Somogyi, P., Cowey, A., Halasz, N., and Freund, T.F. (1981). Vertical organization of neurons accumulating ^{3}H-GABA in visual cortex of rhesus monkey. *Nature 294,* 761–763.

Somogyi, P., Cowey, A., Kisvarday, Z.F., Freund, T.F., and Szentagothai, J. (1983b). Retrograde transport of γ-amino [^{3}H]butyric acid reveals specific interlaminar connections in the striate cortex of monkey. *Proc. Natl. Acad. Sci. USA 80,* 2385–2389.

Somogyi, P., Freund, T.F., Hodgson, A.J., Somogyi, J., Beuroukas, D., and Chubb, I.W. (1985). Identified axo-axonic cells are immunoreactive for GABA in the hippocampus and visual cortex of the cat. *Brain Res. 332,* 143–149.

Somogyi, P., Freund, T.F., and Kisvarday. Z.F. (1984). Different types of ^{3}H-GABA accumulating neurons in the visual cortex of the rat. Characterization by combined autoradiography and Golgi impregnation. *Exp. Brain Res. 54,* 45–56.

Somogyi, P., Kisvarday, Z.F., Martin, K.A.C., and Whitteridge, D. (1983a). Synaptic connections of morphologically identified and psysiologically characterized large basket cells in the striate cortex of cat. *Neuroscience 10,* 261–294.

Somogyi, P., and Soltész, I. (1986). Immunogold demonstration of GABA in synaptic terminals of intracellularly recorded, horseradish peroxidase-filled basket cells and clutch cells in the cat's visual cortex. *Neuroscience 19,* 1051–1065.

Spitzer, H., Desimone, R., and Moran, J. (1988). Increased attention enhances both behavioral and neuronal performance. *Science 240,* 338–340.

Spitzer, H., and Hochstein, S. (1985). A complex-cell receptive-field model. *J. Neurophysiol. 53,* 1266–1286.

Steriade, M., and Deschênes, M. (1973). Cortical interneurons during sleep and waking in freely moving primates. *Brain Res. 50,* 192–199.

Stone, J., Dreher, B., and Leventhal, A. (1979). Hierarchical and parallel mechanisms in the organization of visual cortex. *Brain Res. Rev. 1,* 345–394.

Stone, J., and Hoffmann, K.P. (1971). Conduction velocity as a parameter in the organisation of the afferent relay in the cat's lateral geniculate nucleus. *Brain Res. 32,* 454–459.

Sur, M., Garraghty, P.E., and Bruce, C.J. (1985). Somatosensory cortex in macaque monkeys: Laminar differences in receptive field size in areas 3b and 1. *Brain Res. 342,* 391–395.

Sur, M., Merzenich, M.M., and Kaas, J.H. (1980). Magnification, receptive-field area, and "hypercolumn" size in areas 3b and 1 of somatosensory cortex in owl monkeys. *J. Neurophysiol. 44,* 295–311.

Sur, M., Wall, J.T., and Kaas, J. (1984). Modular distribution of neurons with slowly adapting and rapidly adapting responses in area 3b of somatosensory cortex in monkeys. *J. Neurophysiol. 51,* 724–743.

Sur, M., Weller, R.E., and Kaas, J.H. (1981). Physiological and anatomical evidence for a discontinuous representation of the trunk in S-I of tree shrews. *J. Comp. Neurol. 201,* 135–147.

Swadlow, H.A., and Weyand, T.G. (1987). Corticogeniculate neurons, corticotectal neurons, and suspected interneurons in visual cortex of awake rabbits: Receptive-field properties, axonal properties, and effects of EEG arousal. *J. Neurophysiol. 57,* 977–1001.

Tanaka, K. (1983). Cross-correlation analysis of geniculostriate neuronal relationships in cats. *J. Neurophysiol. 49,* 1303–1318.

Tasaki, K., Polley, E.H., and Orrego, F. (1954). Action potentials from individual elements in cat geniculate and striate cortex. *J. Neurophysiol. 17,* 454–474.

Toyama, K., Kimura, M., and Tanaka, K. (1981a). Cross-correlation analysis of interneuronal connectivity in cat visual cortex. *J. Neurophysiol. 46,* 191–201.

Toyama, K., Kimura, M., and Tanaka, K. (1981b). Organization of cat visual cortex as investigated by cross-correlation technique. *J. Neurophysiol. 46,* 202–214.

Toyama, K., Maekawa, K., and Takeda, T. (1973). An analysis of neuronal circuitry for two types of visual cortical neurones classified on the basis of their responses to photic stimuli. *Brain Res. 61,* 395–399.

Toyama, K., Maekawa, K., and Takeda, T. (1977). Convergence of retinal inputs onto visual cortical cells: I. A study of the cells monosynaptically excited from the lateral geniculate body. *Brain Res. 137,* 207–220.

Toyama, K., Matsunami, K., Ohno, T., and Tokashiki, S. (1974). An intracellular study of neuronal organization in the visual cortex. *Exp. Brain Res. 21,* 45–66.

Ts'o, D.Y., Gilbert, C.D., and Wiesel, T.N. (1986). Relationships between horizontal interactions and functional architecture in cat striate cortex as revealed by cross-correlation analysis. *J. Neurosci. 6,* 1160–1170.

Tsumoto, T. (1978). Inhibitory and excitatory binocular convergence to visual cortical neurons of the cat. *Brain Res. 159,* 85–97.

Tsumoto, T., Eckart, W., and Creutzfeldt, O.D. (1979). Modifidcation of orientation sensitivity of cat visual cortex neurons by removal of GABA-mediated inhibition. *Exp. Brain Res. 34,* 351–363.

Van Essen, D.C. (1985). Functional organization of primate visual cortex. In: *The Cerebral Cortex.* Vol. 3, Visual Cortex (A. Peters and E.G. Jones, eds.). Plenum Press, New York, pp. 259–329.

Van Essen, D.C., and Kelly, J. (1973). Correlation of cell shape and function in the visual cortex of the cat. *Nature 241,* 403–405.

Welker, C. (1976). Receptive fields of barrels in the somatosensory neocortex of the rat. *J. Comp. Neurol. 166,* 173–189.

Welt, C., Aschoff, J.C., Kameda, K., and Brooks, V.B. (1967). Intracortical organization of cat's motorsensory neurons. In: *Neurophysiological Basis of Normal and Abnormal Motor Activities* (M.D. Yahr and D.P. Purpura, eds.). Raven Press, New York, pp. 255–294.

Weyand, T.G., Malpeli, J.G., Lee, C., and Schwark, H.D. (1986). Cat area 17. III. Response properties and orientation anisotropies of corticotectal cells. *J. Neurophysiol. 56,* 1088–1101.

White, E.L., Benshalom, G., and Hersch, S.M. (1984). Thalamocortical and other synapses of non-spiny multipolar cells in mouse SmI cortex. *J. Comp. Neurol. 229,* 311–320.

White, E.L., and Rock, M.P. (1980). Three dimensional aspects and synaptic relationships of a Golgi impregnated spiny stellate cell reconstructed from serial thin sections. *J. Neurocytol. 9,* 615–636.

White, E.L., and Rock, M.P. (1981). A comparison of thalamocortical and other synaptic inputs to dendrites of two non-spiny neurons in a single barrel of mouse SmI cortex. *J. Comp. Neurol. 195,* 265–277.

Whitsel, B.L., Roppolo, J.R., and Werner, G. (1972). Cortical information processing of stimulus motion on primate skin. *J. Neurophysiol. 35,* 691–717.

Wiesel, T.N., and Hubel, D.H. (1966). Spatial and chromatic interactions in the lateral geniculate body of the rhesus monkey. *J. Neurophysiol. 29,* 1115–1156.

Wilson, P.D., Rowe, M.H., and Stone, J. (1976). Properties of relay cells in the cat's lateral geniculate nucleus: A comparison of W-cells with X- and Y-cells. *J. Neurophysiol. 39,* 1193–1209.

Winer, J.A. (1986). Neurons accumulating [^{3}H] Gamma-aminobutyric acid (GABA) in supragranular layers of cat primary auditory cortex (AI). *Neuroscience 19,* 771–793.

Wolf, W., Hicks, T.P., and Albus, K. (1986). The contribution of GABA-mediated inhibitory mechanisms to visual response properties of neurons in the kitten's striate cortex. *J. Neurosci. 6,* 2779–2795.

Woolsey, C.N. (1964). Cortical localization as defined by evoked potentials and electrical stimulation studies. In: *Cerebral Localization and Organization* (G. Schaltenbr and and C.N. Woolsey, eds.). Univ. of Wisconsin Press, Madison, pp. 17–26.

Woolsey, C.N., Marshall, W.H., and Bard, P. (1942). Representation of cutaneous tactile sensitivity in the cerebral cortex of the monkey as indicated by evoked potentials. *Bull. Johns Hopkins. Hosp. 70,* 399–441.

Wong-Riley, M. (1979). Changes in the visual system of monocularly sutured or enucleated cats demonstrable with cytochrome oxidase histochemistry. *Brain Res. 171,* 11–28.

Woolsey, T.A., and Van der Loos, H. (1970). The structural organization of layer IV in the somatosensory region (SI) of mouse cerebral cortex. *Brain Res. 17,* 205–242.

Yamamoto, T., Samejima, A., and Oka, H. (1987a). Morphological features of layer V pyramidal neurons in the cat parietal cortex: An intracellular HRP study. *J. Comp. Neurol. 265,* 380–390.

Yamamoto, T., Samejima, A., and Oka, H. (1987b). Morphology of layer V pyramidal neurons in the cat somatosensory cortex: An intracellular HRP study. *Brain Res. 437,* 369–374.

Zarzecki, P., Blum, P.S., Bakker, D.A., and Herman, D. (1983). Convergence of sensory inputs upon projection neurons of somatosensory cortex: Vestibular, neck, head, and forelimb inputs. *Exp. Brain Res. 50,* 408–414.

Zarzecki, P., and Wiggins, D.M. (1982). Convergence of sensory inputs upon projection of somatosensory cortex. *Exp. Brain Res. 48,* 28–42.

Part III Facts, Theories, and Models

EDWARD L. WHITE AND ASAF KELLER

Parts I and II have focused on the concept of the brain as a system of neurons interconnected by synapses. Applying this point of view to the anatomy of the cerebral cortex has enabled the identification of certain basic rules by which the cerebral cortex is put together, thereby illuminating some of the constraints under which it operates. Discussion of the functional aspects of cortical neurons provided additional evidence for the operation of the cerebral cortex according to these and to other constraints. Much of the information presented in Parts I and II is of very recent origin and has yet to be integrated into current theories of brain function. The purpose of Part III is to facilitate this process of integration by examining certain current notions of cortical structure and proposals for cortical function, with the general aim of modifying them to conform to the data and conclusions presented in Parts I and II.

An Integrative View of Cortical Circuitry

7

Synaptic Triads

The data provided in Parts I and II are fully consistent with the notion that the cerebral cortex is organized along similar lines in different species. Lamination and vertical connections between laminae are hallmarks of all cortical systems, the morphological and physiological characteristics of cortical neurons are equivalent in different species, as are the kinds of synaptic interactions involving cortical neurons. This similarity in the organization of the cerebral cortex extends even to the specific details of cortical circuitry. For example, in all cortical areas and in all species the cell bodies of spiny neurons receive input only from nonspiny neurons, whereas the cell bodies of nonspiny neurons receive, in addition, inputs from a variety of other sources (see Peters, 1987).

The similarities in the organization of the cerebral cortex in different species is the basis for our proposition that the rules for cortical organization set out in Chapters 3 and 6 apply to all areas of the cortex in all species. Whether these rules apply to the organization of synaptic circuitry in other regions of the brain is unknown. However, the fact that at least one particular type of synaptic circuit is present in several different regions of the brain, including the cerebral cortex, implies that the brain may well operate under a common set of rules.

Lund (1969) applied the word *triplet* to a circuit in the superior colliculus of the rat that consists of synapses from retinal afferents onto dendrites of both projection neurons and intrinsic neurons. In addition,

Recent research efforts of E.L. White and E. Elhanany were supported by grants from The National Institutes of Health (Grant NS 20149) and from The U.S.–Israel Binational Science Foundation (Grant 86-00041).

dendrites belonging to the intrinsic neurons synapse onto dendrites belonging to the projection neurons. Comparable sequences of axodendritic and dendrodendritic synapses, often including a return dendrodendritic synapse from the projection to the intrinsic neuron, have been shown to occur in other regions of the brain: In the lateral geniculate nucleus of the cat (Famiglietti, 1970), where these synaptic sequences have been termed a *triadic relationship* (Famiglietti and Peters, 1972), in the retina of the frog, pigeon, cat, and primate (Dowling, 1968; Dowling and Boycott, 1965, 1966; Kidd, 1962; Raviola and Raviola, 1970), in the olfactory bulb of the mouse, rat, and rabbit (Andres, 1970; Pinching and Powell, 1971; Reese and Brightman, 1970; White, 1972, 1973), in the medial geniculate body (Morest, 1971), and in the ventral posterior nucleus (Jones and Powell, 1969; Ralston and Herman, 1969) of the cat. In each of these regions, the triadic elements that receive afferent input are dendrites, which interact by means of dendrodendritic synapses. Synapses from the projection cell dendrites to the intrinsic cell dendrites are asymmetrical and presumed excitatory; return synapses from the intrinsic cell dendrites are symmetrical and presumed inhibitory. For a discussion of various functional aspects of these and of other triadic arrangements, see Shepherd (1979).

Triadic arrangements also occur in the cerebral cortex, where a common pattern involves ascending input to excitatory and inhibitory neurons, and reciprocal synaptic interactions between them (Chapter 5). An important difference between triadic arrangements in the cerebral cortex and those in other regions of the brain is that synaptic interactions of cortical triads are mediated by axon terminals and not by dendrodendritic synapses. Synaptic interactions between two neurons that are mediated by dendrodendritic synapses can be expected to involve graded and immediate postsynaptic effects. In contrast, synaptic interactions mediated by axon terminals depend on the all-or-nothing firing of the presynaptic neuron and may involve comparatively longer transmission times through axons.

The ubiquity of triadic circuits indicates that different regions of the brain in diverse species are characterized by certain basic patterns of synaptic connectivity. This may mean that the rules proposed for the organization of cortical circuitry apply also to the synaptic organization of other regions of the brain.

Hierarchical Versus Parallel Processing

More than 25 years ago, Hubel and Wiesel (1962, 1968) proposed a hierarchical model of cortical function in which input from the thalamus enters the primary visual cortex (area 17) and is then passed in stepwise fashion through a series of cortical neurons having progressively more "complex" receptive field properties (see Chapter 4). The basic premise in this scheme is that the receptive field properties of cortical neurons at the initial stage of the hierarchy are shaped by input from thalamic afferents, whereas the response properties of neurons at subsequent stages of the hierarchy result primarily from the convergence of input from neurons with simple receptive field properties onto neurons with more complex ones.

In an attempt to explain their hypothesis in terms of the anatomy of the cerebral cortex, Hubel and Wiesel proposed that the flow of input from the thalamus is first to neurons having somata in layer IV, then from layer IV to the more superficial layers of the cortex, and finally to the deeper layers containing the somata of large pyramidal cells, which then act to send the "finished product" to another region of the brain. At the time these proposals were made, virtually nothing was known of the details of synaptic connections within the cortex; however, there seemed to be sufficient evidence from light microscopic studies to support the existence of a set of afferent and intracortical projections that could serve as the anatomical substrate for the proposed feed-forward pathway.

The hypothesis of convergent processing by a hierarchically organized set of cortical neurons was extended by Hubel and Wiesel (1965, 1977) to provide a description of the functional relationships between different areas of visual representation in the cerebral cortex. At that time, the areas of cortex most commonly associated with visual processing in higher animals were the cytoarchitectonic areas 17, 18, and 19; each area was proposed to perform, in turn, progressively higher levels of visual information processing. It is interesting to note that the designation of these areas, many years earlier, by a sequence of integers increasing by one, was fortuitous and made without regard to the functional organization of the areas or to the functional relationships between them (Brodmann, 1909). The notion that cortical visual areas are hierarchically arranged was subsequently extended by Van Essen (1985) to include a seven-level hierarchy composed of 17 separate visual areas in the monkey; most of these areas were identified following the promulgation of Hubel and Wiesel's model.

Features of cortical organization consistent with the notion that thalamic input is processed by a system of convergent, hierarchical arrangements are (1) the high concentration within layer IV (e.g., in the primary visual cortex [Gilbert, 1977; Hubel and Wiesel, 1968, 1977; and see Figure 4.1] and in the primary somatosensory cortex [Chapin, 1986; Ito, 1985; Simons, 1978; Simons and Woolsey, 1979] of units whose receptive fields are related to those of thalamocortical neurons, and the existence within this same layer of extensive ramifications of thalamocortical afferents (Part I); (2) the high concentration in layer III of complex units (Gilbert 1977; Hubel and Wiesel, 1968, 1977; Simons, 1978; Simons and Woolsey, 1979) coupled with the demonstration of an axonal projection from neurons in layer IV to layer III (Figure 2.2, and Chapter 2, "Collateral Axonal Branches"); (3) the presence in layer V of large numbers of units whose receptive fields may be explained by a convergence of input from units in layer III (Gilbert 1977; Hubel and Wiesel, 1968, 1977; Ito, 1985; Simons, 1978, 1985; Simons and Woolsey, 1979) coupled with an extensive intracortical projection from layer III to layer V neurons (Figure 2.2, and Chapter 2, "Collateral Axonal Branches"); and, finally, (4) regarding interareal relationships, the progressively higher proportion of units having more complex receptive field properties in areas 17, 18, and 19, respectively (Hubel and Wiesel, 1965). Differences in the functional properties of neurons in multiple areas of visual representation are grossly oversimplified in point (4). For a more up-to-date and in-depth comparison of the functional properties of neurons in different visual areas, the reader is referred to Van Essen (1985, pp. 313–320).

The high concentration of simple cells in layer IV where nonpyramidal neurons abound, and of complex and hypercomplex cells in layers above and below layer IV where pyramidal neurons predominate, prompted Hubel and Wiesel (1962, 1968) and later Gilbert (1977) to equate simple units with nonpyramidal neurons and the more complex units with pyramidal neurons. Subsequent work has provided only partial support for these suppositions; at least in the visual cortex where structure/function correlations have been pursued most intensively, a strict correlation between neuronal morphology and receptive field type has not been found (see the discussion in Chapter 4).

Hubel and Wiesel's proposals for the functional organization of the cerebral cortex revolutionized our concept of how the brain processes the information it receives, and much of what they proposed a quarter of a century ago is still being used today to describe the functional organization not only of the visual areas, but also of areas related to other sensory systems. For example, hierarchical processing arrangements have been proposed to explain the functional organization of neurons in somatosensory areas I and II of the monkey (Pons et al., 1987) and in the vibrissal region of somatosensory area I in the rat (Simons, 1978, 1985) and mouse (Simons and Woolsey, 1979).

Despite the popularity of hierarchical processing schemes, it should be recognized that any concept of the hierarchical processing of information by the cortex must be tempered by a consideration of knowledge gained since this concept was first applied by Hubel and Wiesel. For instance, subsequent investigation has shown that the response properties of neurons in the visual cortex are far more elaborate than originally thought (see the discussion in Chapter 4). These findings have provoked criticism of the hierarchical model on the grounds that the criteria employed to classify cortical neurons ignore many features of their response properties and that, for this reason, the classification system employed is arbitrary (see Henry, 1977). Employing a classification system that includes some response properties of cortical neurons but not others casts doubt on the ability of this model to predict the sequence of information processing between functionally identified neurons.

In Hubel and Wiesel's original thesis (see 1977) and in subsequent applications of it to explain findings both in the primary visual (e.g., Gilbert and Wiesel, 1979) and primary somatosensory cortex (Simons, 1978, 1985; Simons and Woolsey, 1979), considerable emphasis was placed on the orderly progression of synaptic activity from neurons whose somata occur in layer IV, through neurons having somata in layers III, to neurons having somata in layer V. Axonal pathways beginning with cell bodies in layer IV and terminating in layer III, and beginning with cell bodies in layer III and terminating in layer V, have been illustrated in studies of neurons labeled by the Golgi method (e.g., Lorente de Nó, 1922; Lund et al., 1979, 1981), or by the intracellular injection of HRP (e.g., Gilbert and Wiesel, 1981), and examined with the light microscope. However, as pointed out in the summary for Chapter 2, information obtained with the light microscope is inadequate to identify the pre- and postsynaptic elements involved in synapses and, in fact, most of the key steps in the proposed hierarchical pathway have not been demonstrated unequivocally.

The crux of the matter is simply that knowing within which layer of the cortex a projection terminates does not provide any specific information regarding the identity of the neuronal type or types that are postsynaptic to these afferents. Unfortunately, the literature contains numerous instances where, in disregard of this simple caveat, a projection to a cortical layer is presumed to terminate primarily onto neurons having somata within that layer. Such presumptions fail to take into account the fact that each layer of the cortex contains a multitude of potential postsynaptic elements whose parent somata occur in other layers. According to Rule 1 of Chapter 3, each potential postsynaptic element receives synapses from the projection. Moreover, without direct knowledge of the numbers, distribution, and physiology of these and of other synapses, it is impossible to determine how activity in the projection will affect the activity of different postsynaptic elements.

The problem of making unwarranted assumptions on the identity of postsynaptic elements, and on the degrees to which specific inputs affect them, is nowhere more evident than in the controversy surrounding the identification of the targets of specific thalamic afferents in the fourth layer of the primary sensory cortices. This issue was touched upon briefly in Part I, but bears reconsideration now because of its obvious relevance to hierarchical arrangements that presume as their starting point the site of entry of thalamic input to the cortex.

It has long been known that layer IV of the primary sensory regions of the cerebral cortex is the main terminus for specific thalamocortical afferents and that this layer contains, in addition, a high concentation of small stellate or nonpyramidal neurons (see the review by White, 1979). On this basis the presumption has been made that the principal targets of thalamocortical afferents are nonpyramidal neurons having cell bodies in layer IV and in the immediately supra-adjacent part of layer III (e.g., Eccles, 1984, p. 13; Gilbert and Wiesel, 1981, Fig. 12; Kandel 1985, Fig. 29-10; Szentagothai, 1979). Support for this hypothesis has been obtained from light microscopic analyses of Golgi-impregnated neurons from the cortices of sensory deprived animals, or from animals that had sustained damage at some point along the ascending sensory pathway. For example, the dendrites of layer IV nonpyramidal cells in the visual cortex of cats reared in the dark are fewer, shorter, and more varied in length than the dendrites of layer IV neurons in control preparations (Coleman and Riesen, 1968; Globus and Scheibel, 1967). The dendrites of layer IV nonpyramidal cells typically ramify within layer IV of the visual cortex, but in mice enucleated at birth, these dendrites are oriented away from layer IV to adjacent layers (Valverde, 1968). Finally, dendrites belonging to nonpyramidal cells in layers III and IV of the auditory cortex of rabbits deafened at birth, display abnormal orientations and are longer than similar dendrites of littermate controls (McMullen et al., 1988).

Results of investigations using electron microscopy or electrophysiological recording have provided incontrovertible evidence that thalamocortical afferents to layer IV synapse not only with layer IV nonpyramidal neurons, but also with a wide variety of both pyramidal and nonpyramidal neuronal types whose cell bodies occur throughout layers II to VI (see the reviews by White, 1981, 1986, 1987; and Tables 3.2 and 5.1). Moreover, results of quantitative studies of thalamocortical synapses show that certain nonpyramidal neurons in layer IV, particularly of the nonspiny type, receive very small numbers of thalamocortical synapses (e.g., White et al., 1984; and see Figure 3.8). In contrast, some neuronal types that have cell bodies outside of layer IV, such as corticothalamic projection cells (somata in layers V and VI), receive relatively large numbers of thalamocortical synapses that, in some instances, represent a significant proportion of the excitatory input to these cells (White and Hersch, 1982; Figure 3.8). The possibility that corticothalamic neurons receive a greater share of thalamocortical synapses than that received by layer IV nonpyramidal neuronal types has not been determined, but this possibility cannot be excluded on the basis of available data. In any event, the presence in layer IV of large numbers of dendrites belonging to corticothalamic neurons, and to other nonlayer IV neuronal types that receive large numbers of thalamocortical synapses, indicates that neuronal types outside of layer IV receive a substantial portion of the thalamocortical input to this layer. These findings necessitate the reevaluation of hierarchical schemes that propose as their starting point thalamocortical synapses onto layer IV neurons.

Inherent in the hierarchical processing scheme is the notion that cells with the simplest" receptive field properties (e.g., simple cells in the visual cortex [Gilbert and Wiesel, 1979] and single-whisker units in somatosensory cortex [Simons, 1978] are the exclusive or primary recipients of thalamocortical input. According to this line of reasoning, it could be proposed that corticothalamic neurons that receive large numbers of thalamocortical synapses belong to the complement of simple cells known to occur within layer VI. However, there is an alternative interpretation: It may be that thalamocortical synapses onto corticothalamic neurons are an example of direct thalamocortical input to complex units. Complex and hypercomplex cells may be monosynaptically activated by thalamocortical input (e.g., Hoffman and Stone, 1971; Toyama et al., 1977; and see Table 5.1), and there are complex units that project to the thalamus (Gilbert, 1977; Harvey, 1980; and Chapter 4). Findings showing that virtually every neuron with a dendrite in layer IV receives some thalamocortical synapses are consistent with the idea that neurons of every receptive field type receive some portion of their input directly from the thalamus.

Having thalamocortical synapses onto complex cells would seem to contradict an obvious corollary of the hierarchical scheme, namely, that direct thalamocortical input enters the hierarchy only at the lowest level and is therefore excluded from direct impingement on higher order neurons. However, that a neuron receives thalamocortical synapses means only that it has the possibility of responding to a thalamic volley; it does not have to respond always, nor even frequently, to thalamic input if the numbers of synapses involved, their spatial distribution, the timing of synaptic events, or certain biophysical properties such as the resistances of the cell membrane (Rm) or of the internal milieu (Ri) are such as to prevent or limit such a response. Taking these considerations into account is useful for resolving the apparent discrepancy between the demands of the hierarchical scheme and the likelihood that thalamic afferents synapse onto neurons at all levels of the hierarchy.

In this regard, it may be helpful, if not entirely accurate, to simplify matters by making the following presumptions: One is that simple cells, whose response properties closely follow thalamic activity, receive a large proportion of their input directly from the thalamus. Cells identified as complex have responses that follow thalamic activity less closely, and on this basis the prediction could be made that their major inputs are from sources other than the thalamus. Findings by Singer et al. (1975) showing that simple cells are driven almost exclusively by thalamic input, but that complex cells may be driven both by thalamic input and by inputs from a variety of other extrinsic and intrinsic sources, are consistent with these arguments. According to this proposal, neurons that receive a large proportion of their input from the thalamus have simple response properties, whereas neurons receiving progressively less of their input from the thalamus show progressively more complex response properties. Conceivably, thalamic input to these latter neurons may operate only at subthreshold levels (cf. Zarzecki, 1985). The value of the proposed arrangement lies in its ability to reconcile the hierarchical model with findings showing complex cells to receive thalamocortical input; the idea is that thalamic input may directly affect the physiology of both simple and complex neurons, but that these effects influence the latter less often and to a lesser degree.

The validity of this scheme could be determined by correlating the response properties of neurons with the numbers of thalamocortical synapses they form. This could be achieved by applying quantitative anatomical approaches to examine the distribution of thalamocortical synapses onto neurons whose electrophysiological properties have been recorded. To be of greatest value, these studies will have to take into account not only the numbers of thalamocortical synapses formed, but also other factors that might affect the efficacy of thalamocortical synaptic input, such as the proportion of thalamocortical synapses formed and the spatial distribution of these synapses on the surface of the postsynaptic neuron.

Included in the hierarchical scheme for cortical processing is the notion that synaptic activity flows sequentially from units in layer III to those in layer V (e.g., Hubel and Wiesel, 1977). The predominant neuronal type in both these layers is the pyramidal cell (Peters et al., 1985), and indeed, pyramidal cells whose somata occur in layer III do have axons that project to the deeper cortical layers (Chapter 2, "Collateral Axonal Branches"). However, for reasons made clear above, an axonal projection from layer III pyramids to layer V does not necessarily mean that layer V pyramidal neurons are the sole, or even the primary, target for the projection. However, if such a connection does exist, it must do so under certain anatomical constraints. For example, it is certain that the postsynaptic element of these synapses cannot be the cell bodies of layer V pyramids; the cell bodies of pyramidal neurons receive only symmetrical synapses, whereas the axon terminals of pyramidal neurons are presynaptic only at asymmetrical synapses. Dendritic spines receive asymmetrical synapses, and so it is conceivable that spines belonging to the apical or basal dendrites of layer V pyramids are targets of the axonal projection from layer III pyramids to layer V. Nevertheless, clear evidence of such a connection is lacking, and though it is likely to exist, projections from layer III pyramids to nonpyramidal neuronal types in layer V cannot be excluded. Moreover, some of the units in layer III that, according to the hierarchical model, are considered to project to the deeper layers may be nonpyramidal neurons whose axons form symmetrical, presumed inhibitory, synapses. This type of arrangement would be inconsistent with the notion of a hierarchy composed of feed-forward, excitatory pathways.

As originally proposed, the hierarchical model did not take into account the importance of inhibition for the processing of information by the cerebral cortex. It is known now that inhibitory mechanisms play a significant part in shaping the receptive fields of cortical neurons (see the discussion in Chapter 6). For example, GABA-mediated inhibition is involved in generating the orientation and direction selectivities of many neurons in the visual cortex (e.g., Sillito, 1984; and cf. Simons, 1985, for a discussion of similar phenomena in the somatosensory cortex) and in controlling the size of receptive fields of neurons in the somatosensory cortex (e.g., Dykes et al., 1984). The pivotal role of inhibitory interactions in shaping the receptive field properties of cortical neurons requires the inclusion of inhibitory mechanisms in any model of cortical processing.

Another problem with the original hierarchical model that is related to its emphasis on excitatory, feed-forward pathways is that it does not take into account that alongside every feed-forward pathway exists another pathway of opposite direction and, for all that is known, of equal efficacy. For example, it is true that a pathway from layer III pyramids terminates in layer V, but it is equally true that layer V pyramids have axon collaterals that terminate in layer III where these axon terminals may synapse onto layer III pyramids. With regard to the hierarchical model, the existence of such feedback pathways would imply that neurons at higher levels of the hierarchy project to neurons at lower levels. In fact, there is some evidence that suggests that complex cells feed back onto simple cells: The responses of simple cells in the cat striate cortex are facilitated by stimuli that drive only complex cells (Hammond and MacKay, 1981).

The pathways that link the different cortical areas involved in the processing of visual information are connected by reciprocal connections that are usually interpreted as feed-forward and feedback pathways (e.g., VI [area 17] projects to and receives projections from VII [area 18]). One possible role in visual processing for feedback pathways is the enhancement of parts of an image having common features (Barlow, 1981), but it is also possible that feedback pathways between cortical areas function within the context of the interdependent and perhaps nonhierarchical, processing of visual input within the separate cortical areas (e.g., see Ballard et al., 1983, and the discussion on parallel processing below). In the latter instance, the distinction between feedback and feed-forward pathways is somewhat unclear, and it may be better to label the connections reciprocal so that no polarity of transmission is implied.

As outlined in Chapter 1, Van Essen and Maunsell (1980) used anatomical criteria to assign feed-forward and feedback polarities to the components of the reciprocal pathways that link multiple visual areas in the monkey. Based on the polarity of these connections, a hierarchical ordering of visual areas was proposed, which Van Essen (1985) developed subsequently into a seven-level hierarchical scheme that included 17 separate visual areas. Moreover, Van Essen provides a number of instances in which the functional attributes of neurons within a visual area are consistent with the hierarchical level assigned to that area. The general applicablility of this hierarchical scheme, or of the considerations on which it is based, is somewhat questionable in view of the difficulties encountered in developing similar schemes for the processing of visual information in other species. For example, Symonds and Rosenquist (1984) were unable to develop an unambiguous hierarchical scheme for the multiple visual areas in the cerebral cortex of the cat using approaches similar to those employed for the monkey (and see Rosenquist, 1985). However, it cannot be excluded that, in light of additional findings, a hierarchical scheme will be developed for the multiple visual areas of the cat, or that the scheme proposed for the monkey will prove invalid.

The notion of hierarchical processing mechanisms has found expression in the relationship between cortical neurons having different types of receptive fields within a single visual area and in the relationships between different cortical areas of visual representation. The existence of either of these hierarchical relationships is independent of the other, and they may involve different processes. Hierarchical arrangements between the different laminae in the primary visual cortex have also been proposed on the basis of the patterns of connections that link the laminae with one another and with other parts of the brain (Chapin, 1986; Van Essen, 1985).

The basic idea in all these hierarchical arrangements is that input to the cortex from the ascending sensory pathways passes in stepwise fashion through a series of distinct stations whose degree of sophistication or complexity increases at each level. One problem with these arrangements as applied to the cerebral cortex is that, for the most part, they fail to take into account that the richness of the circuitry that links different neurons, laminae, and areas of the cortex with one another, undoubtedly forms the substrate for a high degree of cooperation between these different cortical entities. The consequent interdependence of these entities causes some conceptual difficulties for hierarchical models that tend to ignore the possibility that entities at lower levels in the hierarchy may be dependent on direct or indirect feedback from higher levels to function properly. Another conceptual problem regarding hierarchical arrangements in the cortex is that each concentrates on thalamic input as its starting point. Because of this, the impression is given that what comes out of the cortex represents a simple stepwise transformation of what went in and that the cortex is in some way inactive until input arrives from the periphery. Again, the richness of cortical circuitry alone provides sufficient basis upon which to refute these impressions.

In a comprehensive review of electrophysiological and anatomical studies in visual cortex, Stone et al. (1979) concluded that the hierarchical model of cortical processing is incompatible with much of the available data and that a model in which different aspects of visual input are processed by separate, parallel channels provides a much more accurate picture of how visual processing proceeds within the cortex. Similar views were adopted by Lennie (1980), but his position departs from Stone's by its greater emphasis on the convergence of parallel pathways onto units that process input in a hierarchically organized framework. Later, Dykes (1983) extended the parallel model to include the processing of somatosensory stimuli by spatially segregated, modality-specific parallel channels that lead from peripheral receptors through subcortical stations, and on to the primary somatosensory cortex, in which the spatial segregation of submodality specific neuronal populations is maintained. For ease of comparison with the above discussion of hierarchical models, the following description will focus on parallel models as they apply to the visual cortex.

Two major schemes have been proposed to describe parallel processing of visual information (see Chapter 5): the X/Y/W model, which has been applied to results in both cats and monkeys (see the reviews by Stone et al., 1979; Lennie, 1980) and a different parallel model developed by Livingstone and Hubel (1987; Hubel and Livingstone, 1987), which as yet has been applied only to findings in the primate visual system.

Evidence for the X/Y/W system includes the existence within the retina of at least three distinct types of ganglion cell (X, Y, and W) that differ in their receptive field properties and morphologies. Axons of these cells project to their counterparts in the thalamus, which in turn project to the primary visual cortex, again with little or no overlap. The terminal fields of X and Y thalamocortical afferents may overlap within the same layers of the primary visual cortex (e.g., Humphrey et al., 1985; and see the discussion in Chapter 5), where each of these afferent types may preferentially innervate specific types of cortical neurons (e.g., Freund et al., 1985). Finally, visual areas 17, 18, and 19 each receive a characteristic mix of afferent types from the thalamus:

Area 17 receives X and Y inputs, area 18 receives its principal input from Y-type afferents, and area 19 from W-type afferents, probably from the superior colliculus (e.g., Dreher et al., 1980). Anatomical data suggesting that nearly every neuron in the cerebral cortex receives some portion of its input directly from the thalamus (White, 1987) and that neurons in nearly every layer of the cortex are monosynaptically activated by thalamic afferents (see Martin, 1984, Table 5.1) provide further evidence in support of the notion that thalamocortical input is processed along parallel channels.

Livingstone and Hubel's (1987) model provides anatomical, physiological, and psychophysical evidence for the existence of separate channels for processing visual information and, in addition, assigns each channel a specific role for processing and transmitting a different aspect of visual perception (see Figure 5.2). Evidence in support of the existence of these parallel information channels includes (1) the existence of parallel anatomical pathways, originating in the retina and projecting through the thalamus to the primary visual cortex and to other cortical visual areas; (2) the segregation of cells belonging to the different informational channels within specific cortical areas and within cortical regions defined by cytochrome oxidase staining; (3) the similarities between the receptive field properties of neurons within each of these channels—each channel contains cells that respond preferentially to particular stimulus parameters; and (4) results of psychophysical and clinical studies in humans. According to this model, the "magno" channel is responsible for analyzing depth, movement, and form perception, the "blob" system is involved with color perception, and the "parvo-interblob" channel deals with fine analysis of form perception (see the discussion in Chapter 5).

The two models of parallel processing have in common that they take into account both anatomical and physiological evidence for the segregation of specific neuronal types at several stages of visual information processing (i.e., in the retina, thalamus, and cortex). However, the X/Y/W model describes parallel processing between neurons within each channel that have similar firing properties, whereas the magno/parvo model describes the processing along each channel of different aspects of visual information (form, color, etc.). At present, the relationship between the sets of parallel pathways described by the two different models is unclear, but there is no reason to believe that the models are mutually exclusive. For example, "the blob channel" can be envisioned to contain neurons that belong to the X or to the Y pathways.

Stone et al. (1979) and Dykes (1983) mention a number of instances in which the parallel model is far better suited than the hierarchical one to explain certain psychophysical phenomena of vision and somatosensory sensation (cf. Carney et al., 1987; Livingstone and Hubel, 1987), and Ballard et al. (1983) have demonstrated the appropriateness and efficacy of applying parallel processing algorithms to the development of computational models for visual processing. Although certain aspects of information processing by the brain may be more adequately explained by the parallel or by the hierarchical model, in principle, the two models are compatible, and the possibility that both processing modes operate interactively as well as separately within the cortex cannot be excluded. For example, in accord with a hierarchical arrangement, Ito (1985) reports that laminar differences in various response properties of units in the rat primary somatosensory cortex are most apparent when units in layer IV are compared with units in layer Vb, "the input and output stages of the cortex, respectively." However, measures of response latencies show no significant differences between these presumably low- and high-level hierarchical units, a finding consistent with the contention (White, 1979) that neurons throughout the full thickness of the neocortex receive input, in parallel, directly from the thalamus.

Alternatively, it may be that parallel operations predominate during the initial stages of cortical processing, after which they are replaced by hierarchical mechanisms. For example, as Dow (1978) points out, the parallel scheme essentially purports the existence of a set of neurons that integrate and analyze separate aspects of the peripheral stimulus (i.e., feature detectors); some higher level of cortical processing must of necessity be invoked to explain the reassembling of the many separate bits of information into a coherent picture of the environment.

The different channels of visual information processing proposed by Livingstone and Hubel (1987) may be viewed as neuronal networks that integrate and analyze separate aspects of visual stimuli. Neurons that combine the information provided by these integration networks would display the kinds of highly specific preferences for stimuli that are considered characteristic of neurons at higher levels of a processing hierarchy.

An intriguing possibility is that the brain employs parallel modes for only certain aspects of information processing and integrates the results into what is basically a sequential, hierarchical framework (see Palm and Bonhoeffer, 1984, for applications of this approach in computer science). An example of this arrangement is found in the auditory system of the bat where different physical characteristics of biosonar information are processed initially in parallel by separate parallel channels originating in the cochlea (Suga, 1984, 1987). In the medial geniculate nucleus, information from the separate channels is integrated to produce a more limited set of parallel channels, each dealing with a different aspect of the biosonar information, such as the range and velocity of the prey. In the auditory cortex, some neurons respond to a combination of several elements of biosonar information, indicating a further convergence of parallel channels at the level of the cerebral cortex (Suga, 1984; Suga and Horikawa, 1986). Thus, in the bat auditory system, information is carried along parallel channels that coexist within a hierarchical processing framework.

In summary, although the action and interaction of hierarchical and parallel processing modes remains a matter of speculation, their coexistence does not of itself present any serious conceptual problem. In one sense, the parallel and hierarchical models resemble each other quite closely: Each model takes as given that the response properties of cortical neurons are fixed—units are not at some times simple and at other times complex, and neurons most responsive to one type of input (e.g., parvocellular or magnocellular) do not suddenly respond preferentially to another type of input. These features are based on experimental findings that (1) reflect the stability in the anatomical organization of synaptic pathways in the normal adult brain, and (2) provide support for the notion that the electrophysiological responses of single units are determined by the specific set of connections impinging on them.

Specificity of Synaptic Connections

Investigation of the specificity of synaptic relationships between pre- and postsynaptic elements has tended to follow two basic lines of inquiry: One concerns the propensity of identified postsynaptic elements to form synapses of a particular type, a line of inquiry exemplified by quantitative studies of thalamocortical synapses with identified neurons. The results of these studies were formulated into the rules that appear at the end of Chapter 3.

The second line of inquiry has a complementary focus, seeking to determine the degree of selectivity exhibited by sets of axon terminals for their postsynaptic elements; data of this type are summarized by the corollaries to the rules listed in Chapter 3. Both lines of inquiry require the generation of large amounts of data concerning the precise numbers, types, and locations of synapses formed between identified neuronal elements.

Most quantitative data on the synaptic connections of identified elements have been acquired within only the past few years. As a consequence, quantitative studies of identified synapses are available for but a few areas of the cerebral cortex and in a very small number of species. Nevertheless, having such fresh and exciting data at hand makes it all too inviting to draw comparisons between, for example, the local output relationships of neurons in one part of cortex or in one species with those of similar neurons in a different part of the cortex or in another species. However, as mentioned earlier in Chapter 3, the validity of interpretations based on such comparisons hinges on the degree of similarity in the composition of the neuropil being compared. In short, it is not enough to know the proportions of the different postsynaptic elements contacted; it is also necessary to determine the concentrations of these elements within the regions under study. Only in this way can it be determined whether the proportions of the different postsynaptic elements contacted by a particular axonal pathway are merely a reflection of the relative concentrations of different postsynaptic elements in the neuropil or, alternatively, whether the axons have sought out and formed synapses only with specific postsynaptic elements.

Evidence that axonal pathways of both extrinsic and intrinsic origin are highly selective with regard to the elements with which they synapse is available at present for two regions of the cerebral cortex: The primary somatosensory cortex of the mouse and the primary visual cortex of the cat. Figures 7.1 and 7.2 compare, for certain laminae in each area, the numbers of all asymmetrical synapses formed in the neuropil onto spines and dendritic shafts, with the numbers of asymmetrical synapses onto spines and shafts formed by identified afferents. Each figure is a synthesis of data culled from several publications and obtained using approaches

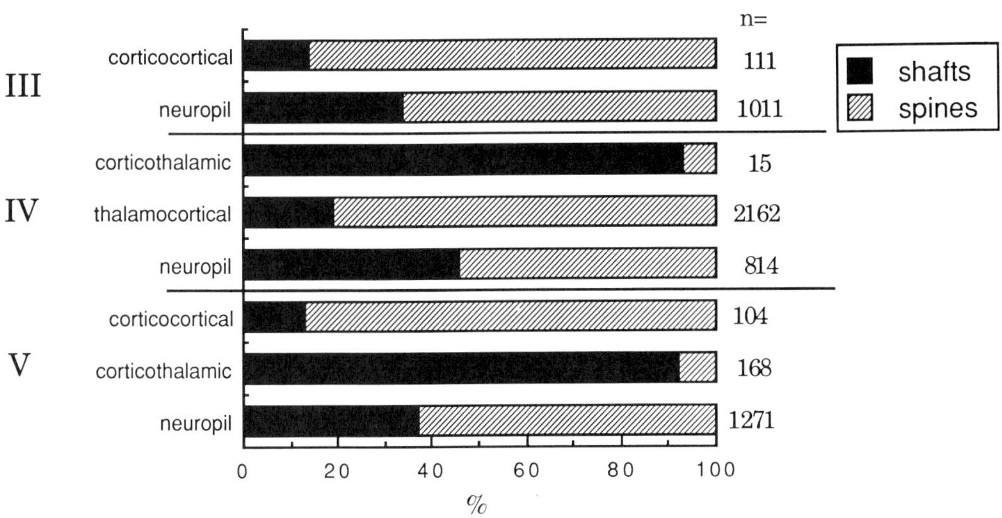

Figure 7.1. *Synapses onto dendritic shafts and spines, layers III, IV, and V somatosensory cortex.* Corticocortical, *asymmetrical synapses made by local axon collaterals of neurons that project to the ipsilateral MsI cortex (Elhanany and White, unpublished data).* Corticothalamic, *asymmetrical synapses made by local axon collaterals of neurons that project to the ipsilateral VB thalamus (White and Keller, 1987).* Thalamocortical, *asymmetrical synapses made by afferents from the ipsilateral VB thalamus (Elhanany and White, unpublished data).* Neuropil, *all asymmetrical synapses onto axons and dendrites in layers III, IV, or V of the mouse SmI cortex (Elhanany and White, unpublished data). Values of $p << 0.0001$ were obtained in Chi^2 tests comparing (1) axodendritic (shaft) versus axospinous corticocortical synapses with all asymmetrical, axodendritic, and axospinous synapses in the layer III neuropil; (2) shaft versus spine thalamocortical and corticothalamic synapses with all asymmetrical, axodendritic, and axospinous synapses in the layer IV neuropil; and (3) shaft versus spine synapses of each pathway terminating in layer V with each other and with the layer V neuropil. Because of low expected values, Fisher's Exact test was performed for comparisons of corticothalamic synapses in layer IV with thalamocortical synapses in this layer; for this comparison, a value of $p \leq 0.0002$ was obtained.*

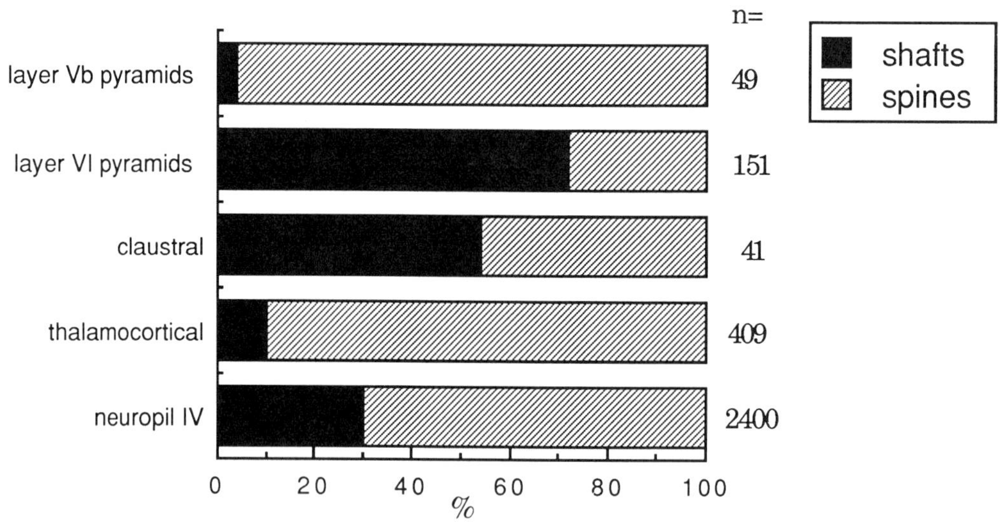

Figure 7.2. *Synapses onto dendritic shafts and spines; layer IV visual cortex.* Layer Vb pyramids, *asymmetrical synapses made by local axon collaterals of two neurons (Gabbott et al., 1987).* Layer VI pyramids, *asymmetrical synapses made by local axon collaterals of two neurons (McGuire et al., 1984).* Claustral, *asymmetrical synapses made by afferents from the ipsilateral claustrum (LeVay, 1986).* Thalamocortical, *asymmetrical synapses made by afferents from the dorsal lateral geniculate nucleus (Garey and Powell, 1971; LeVay, 1986; Winfield and Powell, 1983).* Neuropil, *all asymmetrical synapses onto axons and dendrites in layer IV of the cat primary visual cortex (Beaulieu and Colonnier, 1985; Beaulieu, personal communication).*

Values of $p \leqslant 0.0001$ were obtained in Chi^2 tests comparing axodendritic (shaft) versus axospinous synapses of each projection with all asymmetrical, axodendritic, and axospinous synapses in the layer IV neuropil.

that may not be comparable. For instance, in certain studies, axon terminals were identified by lesion-induced degeneration or by the retrograde transport of horseradish peroxidase. Applied correctly, these methods will label all the axon terminals belonging to a particular axonal pathway. In other studies, axons were labeled by the intracellular injection of horseradish peroxidase, a method that labels only axon terminals belonging to the neuron that was injected.

The data illustrated in Figures 7.1 and 7.2 show that the proportions of spines and dendritic shafts contacted by both intrinsic and extrinsic synaptic pathways differ significantly (see legend) from the proportion of axospinous versus axodendritic synapses within the neuropil of the target regions. In addition, each pathway forms a characteristic proportion of its synapses with spines and shafts. Thus, in all layers in which they occur, the local axon collaterals of corticothalamic projection neurons synapse mainly with dendritic shafts, whereas terminals belonging to the local axon collaterals of corticocortical projection neurons synapse mainly with spines. These findings, which are consistent with the corollaries listed at the end of Chapter 3, are interpreted to reflect the specificity inherent in the synaptic output relationships of both intrinsic and extrinsic axonal pathways. In addition, the findings illustrated in Figure 7.1 provide an indication of the proportions of different neuronal types that are postsynaptic to the labeled pathways. Most of the postsynaptic shafts belong to nonspiny neurons (confirmed by analyses of serial thin sections), and so the proportions of synapses onto shafts can be considered a rough indication of the proportions of synapses onto nonspiny neurons.

Comparisons of the asymmetrical, axodendritic (shaft) versus axospinous synapses in the neuropil (see Figure 7.1) showed significant differences between layers III and IV and between layers IV and V (Chi^2 tests, $p << 0.0001$), but no significant difference between the composition of the neuropil in layers III and V (Chi^2 test, $p << 0.333$). The neuropil in layer IV differs from that in layers III and V by having a slightly higher proportion of asymmetrical synapses onto shafts than onto spines. Presumably, this result reflects a basic difference in the neuronal composition of layer IV vis-à-vis layers III and V. Layers III and V are distinguished by the presence of pyramidal cell bodies, whereas in layer IV nonpyramidal cell bodies are more prominent (Chapter 1, but see Peters et al., 1985). However, these features alone do not indicate whether neuropil in layer IV contains, for instance, a higher proportion of nonspiny dendrites, which are known to form large numbers of asymmetrical synapses, than the neuropil of layers III and V.

Despite the significant difference in the composition of the neuropil in layers IV and V (Fig. 7.1), corticothalamic axon collaterals form the same proportion of shaft versus spine synapses in each layer (Chi^2 = 1.0). This suggests that the characteristic proportion of synapses formed by an axonal pathway is more closely linked to properties inherent in the pathway than to the composition of the neuropil in the target region.

Pathways afferent to motor areas of the cortex also form specific proportions of their synapses (all asymmetrical) with spines and dendritic shafts. For example, callosal afferents to area 4 of the monkey motor cortex form 96% of their synapses with spines; afferents to this region from the primary somatosensory cortex, 82%; and afferents from area 6, 76% (Sloper, 1973; Sloper and Powell, 1979). No information is available in this instance on the synaptic composition of the neuropil. However, the target regions for these afferents overlap within the supragranular layers of the motor cortex. Thus, it would seem that, at least for two of these projections, the proportions of axospinous and axodendritic synapses formed differ from that present within the neuropil. These findings indicate that cortical pathways in motor as well as in sensory regions are highly selective with regard to the elements with which they synapse.

Neuronal systems outside the cerebral cortex also display selectivity in the kinds of synaptic connections they form. For example, in the lateral geniculate nucleus, single retinal afferents are highly selective for the numbers and distribution of the synapses they form (Hamos et al., 1987). Another example is found in the spinal cord where Ia-afferents project preferentially to the alpha motoneurons that innervate their muscles of origin (Burke, 1981). These findings indicate that the occurrence of specific patterns of synaptic connections is a general organizational principle of the nervous system.

Nonselectivity of Synaptic Connections: The Basis for a Selection Process

Anatomical findings showing axonal pathways to synapse onto every element within their target region (Rule 1, Chapter 3) have been interpreted to indicate that cortical wiring is random or nonselective (cf. Peters, 1979). For the simple reason that biological systems usually contain some degree of imprecision, the presence of nonselectivity in an essentially selective system could be expected. However, that is not to say that nonselectivity or randomness in synaptic connections, even if present, need be relegated to the status of background noise. On the contrary, nonselective or random arrangements of synaptic connections may well play an important role in cortical function. To illustrate one form this role might take, we turn now to a model developed from conclusions based on the patterns of synaptic connections formed by thalamocortical afferents.

Thalamic afferents to layer IV of the primary sensory areas of the cerebral cortex terminate on dendrites irrespective of their cell type of origin (Chapters 3 and 5). Should that prove true for thalamic projections to layers I and VI, then it is likely that neurons throughout the full thickness of the neocortex receive input directly from the thalamus. Direct thalamic input to different types of neurons having somata in all the cortical layers suggests that an increase in the rate of firing of thalamic neurons as a result of peripheral stimulation would be detected by, and to some degree directly alter the physiology of, large numbers of neurons throughout the full thickness of the area(s) of the cortex that is topographically related to the excited region(s) of the thalamus. It is reasonable to presume that the firing thresholds of cortical neurons that receive this input will be lowered, or that their firing rate will be increased sufficiently to facilitate the transmission and processing by the affected neurons of input from the thalamus or from other sources. This process, in effect, may be viewed then as a selection mechanism whereby groups of neurons, which might not always act in concert, are, by virtue of common thalamic input, selected out to form a functional unit (cf. White, 1978, 1979, 1981).

According to the "searchlight hypothesis" stated by Crick (1984; and see Sejnowski, 1986), focal inhibition originating in the reticular complex produces bursts of firing in a subset of thalamocortical neurons that are activated by peripheral stimuli. The bursting activity of this group of thalamic neurons, that is, the "searchlights," is thought to produce transient alterations in the efficacy of synapses involving the cortical neurons onto which they synapse. In this way, the cortical neurons are recruited into specific circuits by virtue of their common thalamic input. The selection process we describe and the searchlight hypothesis have in common that an apparently random or nonselective arrangement of thalamocortical synapses could have effects on cortical processing that ultimately, and ironically, are in fact selective.

The selection process need not be restricted to thalamic afferents: Other extrinsic and even intrinsic pathways typically synapse with a variety of postsynaptic elements (Chapter 1), an arrangement that would enable them, also, to recruit groups of cortical neurons for specific tasks. In addition, the proposed recruitment or selection of neurons does not have to be effected by a single axonal pathway; but may act through an intermediary. For example, thalamic stimulation results in catecholamine release in the cat and monkey visual cortex (Marrocco et al., 1987). There is reason to believe that catecholamines, released either at synapses (Silverman et al., 1985) or diffusely into the neuropil (Descarries et al., 1977), may act to enhance signal-to-noise ratios, thereby controlling the overall responsiveness of neurons to synaptic input (Chapter 1). A mechanism can be envisioned for the cerebral cortex in which the selection of a population of neurons is brought about by catecholamine mediated responses consequent to thalamocortical stimulation.

The selection of cortical neurons by input from the thalamus or from other sources is certain to involve GABAergic inhibitory neurons: GABAergic neurons are postsynaptic to thalamocortical afferents (e.g., Keller and White, 1987), and current data indicate that every cortical neuron is subject to GABAergic inhibition (Chapter 6). Presumably, thalamic input activates GABAergic neurons, which in a process of feed-forward inhibition inhibit other neurons. Neurons that are inhibited by this feed-forward pathway would not be able to participate in the circuitry selected by the thalamic volley, whereas neurons that are not inhibited would be able to. GABAergic inhibition can also be expected to exert more subtle effects on postsynaptic neurons since, as we have seen, the integration of inputs by neurons depends to a large extent on spatiotemporal interactions between these inputs (Chapter 6). This suggests that GABAergic inhibition may be involved in the selection of neurons not only by preventing a cell from firing, but also by an intricate process of controlling the relative efficacy of various inputs to the cell. In addition to participating in feed-forward inhibitory processes, GABAergic inhibition may mediate the recruitment of neurons into circuits by selectively removing tonic inhibition from cortical neurons. Tonic inhibition could be removed by inhibiting GABAergic neurons that are activated by specific thalamic input. The disinhibition of some cortical neurons would then enable them to participate in the selected circuit. The pivotal role of GABAergic inhibition in shaping the receptive field properties of cortical neurons (see Chapter 6) is consistent with the hypothesis that GABAergic cells are involved in the process of neuronal selection.

In a series of studies, Merzenich and colleagues (Merzenich and Kaas, 1982; Merzenich et al., 1983, 1984) provide strong evidence for the existence of selection processes in the somatosensory cortex. These investigators report that the topographic map of the hand in the monkey somatosensory cortex is significantly altered following denervation of the hand. Cortical regions previously activated by stimulation of the deafferented body surface immediately become responsive to stimulation of neighboring surfaces of the hand. These rapid alterations in cortical representation cannot be explained by the formation of new synapses, and other possible mechanisms for the plasticity of synaptic interactions need to be invoked. For example, the removal of specific thalamic input to the somatosensory cortex may abolish the feed-forward inhibition mediated by GABAergic neurons. The removal of these inhibitory interactions could reveal latent thalamocortical input to a cortical region previously unresponsive to stimulation from that area of the skin. The finding that removal of GABAergic inhibition results in an increase in the size of receptive fields of neurons in the cat somatosensory cortex (e.g., Hicks and Dykes, 1983; and see Chapter 6) is consistent with this mechanism. Neurons in the rat barrel cortex that are normally unresponsive to deflections of a specific vibrissa can be activated by stimulating this vibrissa following the presentation of a conditioning stimulus that includes the simultaneous deflection of this vibrissa and of others (Delacour et al., 1987). The plasticity in receptive fields revealed by Delacour et al. suggests that latent thalamic input is activated by the paired stimulus paradigm, possibly by the blocking of feed-forward inhibition.

The mechanisms of neuronal selection described above would allow each cortical neuron, which undoubtedly receives many different kinds of inputs and projects to various targets, to be selected for, and thus to participate in, simultaneously, more than one kind of functional unit.

The clearest example of a functional unit composed of cortical neurons are the functional columns (Chapters 1 and 4), and so at this juncture it seems worthwhile to examine the possible relationships between functional columns and the proposed selection mechanisms.

Functional columns have been demonstrated both by physiological and by anatomical methods (Chapters 1 and 4), but this does not mean that an individual functional column is necessarily a static unit composed of a single set of neurons that occupies a particular space within the cortex. Certain columnar structures, such as the barrels in the somatosensory cortex, may fit this definition.

From their study of axonal trajectories between the thalamus and the barrel region of the somatosensory cortex of mice, Bernardo and Woolsey (1987) concluded that thalamic afferents arrive at their appropriate, topographically arranged cortical target areas, that is, within the appropriate barrel, despite the abolition of these topographic relationships at several points along the way. Other studies have provided additional evidence of the intimate relationship between barrels and the thalamic afferents to them (e.g., Killackey et al., 1976, 1978; Woolsey et al., 1981). Barrels, then, may be considered to be columnar structures that are each closely associated with a particular set of afferent inputs. The location and presumably the composition of barrels is similar in different members of the same species. In contrast, the pattern of activity *within* corresponding barrel columns is different in different animals (McCasland and Woolsey, 1988). We concur with McCasland and Woolsey's conclusion that this finding is consistent with the hypothesis that dynamic processes influenced by experience determine patterns of cortical activity.

Other types of columns appear to be far more transitory in nature. These columns owe their existence to the presence of a specific set of stimulus conditions, and it is difficult to associate any one of them with a specific, anatomically definable pathway. Here, in particular, it is important to remember that the connectional strengths between neuronal elements in the cortex reflect dynamic processes that are influenced by experience.

Functional columns that lack clearly defined anatomical substrates are routinely demonstrated by the 2-deoxyglucose method for mapping metabolic activity. Dating from the first application of this method to study the functional organization of the cerebral cortex (Kennedy et al., 1976), the assumption has been that 2-deoxyglucose columns represent collections of neurons whose activity has been increased by the specific experimental conditions employed. Areas between patches of 2-deoxyglucose labeling have been considered to be composed of neurons that are relatively unresponsive to the stimulus, but this may not be entirely true. For instance, Juliano and Whitsel (1987) have demonstrated that neurons in regions of cortex between the 2-deoxyglucose columns may respond to the same stimulus whose presentation is responsible for the formation of the columns.

Crawford et al. (1982) used the 2-deoxyglucose method to identify columns of neurons in the monkey visual cortex that respond preferentially to luminance or to color contrast stimuli. The translaminar shapes of these activity related columns are dependent on properties of the visual stimulus; in different animals, similar stimulus conditions produce similarly shaped columns; different stimulus conditions produce different patterns (Crawford, 1985). Other stimulus conditions produce different activity related columns, such as the ocular dominance and the orientation columns of the visual cortex (see Chapter 4).

Mapping the different systems of functional columns, identified by anatomical and physiological techniques, has resulted in a situation where the visual cortex has more columns than it has space; that is, different columnar systems overlap (see Hubel and Wiesel, 1972, fig 27; Livingstone and Hubel, 1987, fig. 34). One way out of this dilemma is to presume that the formation of functional columns is a transitory phenomenon and so the cortex does not contain the same set of columns from one minute to the next. Because the different columns appear in sequence, the cortex would not at any time contain more columns than it has space.

According to this arrangement, a given nerve cell may at one time belong to one column and at another time to another column. The proposed selection process provides the possiblity for this to happen. A neuron that at one instant fires because it has been activated by one set of inputs may, an instant later, fire as a result of increased activity within a different set of inputs. For example, the auditory cortex of monkeys trained to perform different tasks in response to different auditory stimuli contains neurons that are activated both in response to the auditory stimuli and just prior to the execution of the appropriate motor response (Vaadia et al., 1982). These data indicate that neurons in the cerebral cortex exhibit associative activity in addition to their sensory responses (Abeles, 1982, 1988). Thus, the same neuron may be involved in different functions at different times. In this instance, it is not known if the neurons recruited for one function or the other are arranged in a column, but this is not so important. What is important is to recognize that nerve cells can be selected sequentially for different functional units.

Another way to resolve the problem of the cortex containing more columns or functional units for which it has space is to presume that different columns exist at the same time within the same space. Many neurons in the visual cortex respond preferentially to stimuli having particular orientations, directions of movement, and/or wavelengths (e.g., Gilbert, 1977; Hubel and Wiesel 1962; Michael, 1978a, b). This implies that a neuron that is preferentially responsive to several stimulus parameters may participate simultanously in more than one type of column depending on the composition of the stimulus.

Under normal conditions, an area such as the primary visual cortex responds simultaneously to multiple stimuli, and presumably each set of stimuli produces its own set of columns (cf. Crawford, 1985). Reports using the 2-deoxyglucose method typically present one stimulus at a time, and so only the set of columns related to that stimulus is visualized. This does not mean that other columns do not exist simultaneously, only that current approaches are limited to the visualization of one set of columns at a time. Thus, the demonstration of activity related columns may say more about the conditions of stimulation and recording than about how the visual cortex is organized. Axonal pathways tend to synapse with every element in their terminal field (Corollary to Rule 1, Chapter 1), and neurons receive input from every type of terminal around them (Rule 1, Chapter 1). This kind of mass interaction between contiguous neurons provides a perfect anatomical substrate for intercolumnar interactions that overlap in space and in time.

Selection processes were presented as mechanisms for increasing the activity level of a group of neurons, which by virtue of their reaction to a common stimulus are selected out to form a functional unit. The numbers, types, and distribution of the neurons within the group constantly change according to the stimulus conditions. From this perspective, columns, as elucidated by measures of metabolic activity, may represent visual manifestations of the selection process. In accordance with this idea, we may consider columns to be transitory units whose visualization is a natural outgrowth of the prevalence of vertical connections between cortical laminae (e.g., Blasdel et al, 1985; DeFelipe and Jones, 1985; Lorente de Nó, 1938; Somogyi et al., 1981, 1983).

The hypothesis that neurons in the cerebral cortex are dynamically recruited into functional units forms the basis for a comprehensive theory of cortical function developed by Edelman (1978, 1981). Edelman's group selection theory is particularly valuable because it combines anatomical and physiological data with computer simulations to test the model's applicability to cortical function. The basic premises of the group selection theory are that (1) specific extrinsic and intrinsic cortical connections, developed during ontogeny, provide the basis for the selection of neurons into isofunctional groups or units; (2) these units provide the basis for neuronal representation, or coding of information, and different units interact with each other; and (3) repeated activation of a cortical unit results in alterations in the efficacy of synaptic interactions between members of that unit, such that the probability for the selection of this unit would be increased relative to that of other units.

Pearson et al. (1987) tested the validity of the group selection theory, using a computerized model of the somatosensory cortex. In this model, the "cortex" was designed to receive divergent overlapping extrinsic connections from two types of peripheral receptors, and to contain intrinsic excitatory and inhibitory connections, modeled according to available anatomical data. In addition, the model includes mechanisms for plasticity of synaptic interactions, based on available physiological evidence. Although all the "cortical" cells in this model received essentially the same numbers and types of synapses, repeated afferent stimulation resulted in spontaneously organized neuronal groups. Neurons within each neuronal group responded preferentially to stimuli from one "receptor" type, and the receptive fields of the cells became progressively smaller, stronger, and more selective to the type of stimulus. Decreased stimulation from one set of "receptors," simulating nerve transection, resulted in the reorganization of the "cortical maps"; neuronal groups previously representing the "deafferented" receptors became responsive to stimuli from the other receptor type. Thus, the computerized model, based on the premises of the group selection theory, reproduced the results obtained in studies of the effects of deafferentation on the monkey somatosensory cortex (Merzenich and Kaas, 1982; Merzenich et al., 1983, 1984).

Linsker (1986a, b) constructed a computer model of the visual cortex consisting of a multilayered "cortex" with interlaminar and horizontal synaptic interactions. When the synaptic interactions in this model are endowed with the ability to dynamically alter their strengths, the "neurons" spontanously become orientation specific and are integrated into orientation columns, even in the absence of any extrinsic input to the system. A different model of visual information processing was suggested by Anderson and Van Essen (1987). Their "shifter circuit" model is similar to Edelman's group selection theory in that it postulates that neurons are dynamically selected to participate in different cortical circuits. In this model, diverging excitatory inputs link neurons at different stages of a hierarchically organized system; inhibitory interactions selectively suppress appropriate sets of inputs, thus determining which neurons will participate in the transitory circuits. Anderson and Van Essen propose that their model can be used to solve computational problems in stereopsis, directed visual attention, and the perception of moving images.

In order for cortical neurons to be recruited into more than one cortical circuit either simultaneously or sequentially, these cells must have a massive network of interconnections, for only in this way can the flexibility required for constructing a large number of circuits be achieved. That cortical neurons receive and form synaptic interactions with large numbers of neuronal elements is evident both from anatomical data (Part I) and from statistical analyses of cortical connectivity (e.g., Braitenberg, 1978). At first glance it is difficult to imagine how the cortex would be able to construct well-defined circuits out of this presumably chaotic organization. One possibility is that neurons are selected to perform a certain function if the appropriate stimulus is capable of increasing the firing rates of the cells. These cells would then provide input to a multitude of other neurons, but only those cells whose firing rates are significantly increased by this input would be recruited into the circuit. The recruited cells would proceed to recruit additional cells along the pathway in a similiar manner. Coding of information by the firing rates of neurons may form the basis for hierarchical processing (e.g., Hubel and Wiesel, 1977), in which neurons activated by specific stimuli converge upon, and increase the firing rates of, neurons at higher levels of the hierarchy.

An alternative hypothesis for recruiting neurons into circuits was formulated by Abeles (1982, 1988). According to this hypothesis, information is coded not by the firing rates of neurons, but by the combination of neurons that fire in synchrony. The recruitment of neurons into a circuit does not involve sequential activation of single cells, but rather results from the simultaneous firing of a group of neurons. The chain of neurons that fires in synchrony (i.e., a "synfire chain") transfers information by activating a second synfire chain, which in turn excites another group of simultaneously activated neurons, and so on. The process of recruiting neurons into a synfire chain is a dynamic one; that is, at different times a single cell may participate in different synfire chains. Statistical analyses (Abeles, 1982) have determined that neurons are activated 10 times more efficiently by synchronus presynaptic firing, compared to postsynaptic integration of asychronous inputs. This suggests that coding information by synfire chains has advantages over coding by the firing rates of sequentially activated neurons. In addition, the anatomical basis for synfire chains exists even in a randomly interconnected network of neurons (Abeles, 1988).

The concept of synfire chains includes several attractive properties: It takes into account that large numbers of neurons in the cortex are involved in extensive synaptic interactions; it provides the basis for a practically infinite number of circuits; neuronal death is not likely to affect the activities of the synfire chains; associative processes may be explained by partial overlap in the connections between members of different chains; and finally, fluctuations in the synaptic efficacies of the members of a synfire chain can form the basis for the rapid activation and deactivation of various activities, and thus perhaps for memory and for forgetfulness.

Each neuronal selection process described above requires a mechanism of rapid synaptic plasticity for the formation and dissolution of transient neuronal circuits. The existence of mechanisms by which the efficacy of synaptic interactions can change dramatically was postulated almost a century ago (see discussion in Kandel et al., 1987). This idea received increased attention as a result of the work of Hebb that provided a theoretical framework for plasticity of synaptic interactions. Hebb (1949) hypothesized that "when an axon of cell A is near enough to excite cell B or repeatedly or persistently takes part in firing it, some growth process or metabolic change takes place in one or both cells such that A's efficiency, as one of the cells firing B, is increased."

The concept of the so-called "Hebbian synapse" has attracted wide attention and has undergone several modifications in the last three decades. The most popular postulate for the mechanisms governing plasticity at the Hebbian synapse is that the strength of the synapse is increased if the synapse is activated simultaneously with the spiking of the postsynaptic cell (see the review by DiPrisco, 1984). The wide interest in the concept of the Hebbian synapse has centered mainly on its postulated applicability to memory mechanisms; however, the Hebbian synapse may also be involved in transient neuronal selection processes.

Several pre- and postsynaptic biophysical processes have been proposed as a basis for short-term changes in synaptic efficacy, and experimental results support the existence of some of these mechanism (see the review by Magleby, 1987). For example, Kelso et al. (1986) and Gustafsson et al. (1987) showed that enhancement of synaptic interactions in the hippocampus can be induced by appropriate spatiotemporal interactions between the electrogenic events at the pre- and postsynaptic membranes. This enhancement was shown not to depend on the simultaneous activation of the synapse and firing of the postsynaptic cell, as postulated for the Hebbian mechanism, but rather on the concurrent activation of the synapse with depolarization of the postsynaptic membrane. This so-called pseudo-Hebbian mechanism would provide for the rapid enhancement of synaptic efficacy, a prerequisite for the models of neuronal selection described above. Most research concerning synaptic enhancement has focused on the hippocampus, but recently Bonhoeffer has obtained preliminary indications that synaptic enhancement occurs also in the cerebral cortex (Tobias Bonhoeffer, personal communication). Another recent finding of this investigator indicates that the function of neuronal assemblies may be affected by mechanisms of synaptic enhancement to a far greater extent than hitherto recognized. Following synaptic enhancement by simultaneous stimulation of pre- and postsynaptic elements in hippocampal slice cultures, Bonhoeffer observed the facilitation of the responses of neurons that had not been stimulated previously. These neurons were undoubtedly postsynaptic to terminals of the same axons that had been stimulated as part of the enhancement paradigm, but the recognition that neuronal responses may be facilitated in the absence of prior, intracellular stimulation adds a new dimension to our understanding of the mechanisms of synaptic enhancement.

We can imagine the mechanism of synaptic enhancement applied to the selection of cerebral cortical neurons as follows: Thalamic input will activate large numbers of cortical neurons. On some of these neurons, the postsynaptic membrane will be depolarized at the same time that their thalamocortical synapses are activated, resulting in an increase in the efficacy of these thalamocortical synapses. Cortical neurons receiving this "enhanced" thalamic input would be incorporated into the selected circuit. The dependence of synaptic enhancement on electrogenic activity at the postsynaptic membrane indicates that the efficacy of synaptic interactions is highly dependent on spatiotemporal interactions between various inputs to the postsynaptic cell (see Chapter 6).

In conclusion, even in computer models of neuronal networks where all neurons receive identical patterns of synapses, there are sufficient mechanisms by which neurons can be selected to participate in well-defined circuits. How much easier, then, for a selection process to operate in the cerebral cortex where connections between neurons display a high degree of specificity.

Epilogue

Throughout this work we have recognized as synaptic connections only those that have been identified on the basis of clear, unequivocal evidence. In following a stringent approach, we are likely to have excluded from consideration many intracortical pathways that exist, but for which we consider the evidence inconclusive. Nevertheless, even the "strict" view presented in Figure 3.9 demonstrates that intracortical communication involves transmission over a multitude of pathways, so much so that appropriate pathways can be found to support even seemingly contrasting theories of neuronal interaction. For instance, the circuits in Figure 3.9 allow for both hierarchical and parallel processing within a single group of cortical neurons.

When discussing hierarchical and parallel modes of information processing, we emphasized the likelihood that these mechanisms operate interactively, an argument that presumes the separate existence of both mechanisms. Alternatively, the identification of hierarchical or parallel mechanisms may be related more to a tendency to simplify a complicated system by abstracting individual features from it than to the separate existence of these features. According to this view, the abstracted features, that is, the hierarchical or parallel mechanisms, owe their existence to the abstraction, and so each provides only a limited and misleading description of cortical information processing. In effect, our attention is drawn to the details of the components of the system and thus away from a consideration of the integrative functioning of these components.

The same criticism can be made of attempts to abstract any particular circuit from the wealth of synaptic connections that link cortical neurons with each other and with neurons in other regions of the brain. For example, triadic arrangements may be ubiquitous, but this would be expected in a system where essentially every neuron is connected with each of its neighbors. As Sholl (1956, p. 29) pointed out, in view of the many different circuits in cortex, any scheme relying on specific circuits would be difficult to justify.

We view the cerebral cortex as a system of neurons interconnected by specific patterns of synapses. That the brain is hard-wired and that synaptic connections between its neurons are generated in a specific rather than in a random fashion are concepts that have come in and out of fashion over the years. The wealth of synaptic connections involving each cortical neuron, appears, at least at first glance, to be consistent with descriptions of cortical organization as random or chaotic. However, there is now incontrovertible evidence in support of the specificity of synaptic connections in the cerebral cortex. Moreover, dynamic selection processes enforce an order on synaptic interactions in the cerebral cortex even as they ensure the constant formation and dissolution of specific circuits. As a consequence, individual neurons are sequestered simultaneously or sequentially for different functions. This process makes maximal use of the richness of cortical synaptic connections to provide the system with the highest degree of flexibility.

Despite the recognition that the cerebral cortex is a highly ordered structure, the apparent complexity of its myriad interconnecting pathways provokes despair of ever describing its circuitry and function in any meaningful way. However, in this treatise we have seen that consideration of the numbers of synapses and of their spatial distribution, of the receptive field properties of cortical neurons, and of the types and activities of their neurotransmitters and receptors allows the formulation of meaningful predictions regarding the synaptic organization and function of the cerebral cortex. Thus, there is hope that, as additional information of this type becomes available and as more is discovered about other aspects of cortical function, we will be able to assign appropriate weights to any given set of connections under specific conditions.

References for Part III

Abeles, M. (1982). *Local Cortical Circuits: An Electrophysiological Approach* (Studies of Brain Function, Vol. 6). Springer-Verlag, Berlin.

Abeles, M. (1988). Neural codes for higher brain functions. In: *Information Processing by the Brain: A Physiological-Cognitive Perspective.* (H. Markovitsch, ed.). Hans Huber Press, Berne, pp. 223–232.

Anderson, C.H., and Van Essen, D.C. (1987). Shifter circuits: A computational strategy for dynamic aspects of visual processing. *Proc. Natl. Acad. Sci. USA 84,* 6297–6301.

Andres, K.H. (1970). Anatomy and ultrastructure of the olfactory bulb in fish, amphibians, reptiles, birds and mammals. In: *C.I.B.A. Foundation, Taste and Smell in Vertebrates* (G.E.W. Wolstenholme and J. Knight, eds.). J. and A. Churchill, London, pp. 177–196.

Ballard, D.H., Hinton, G.E., and Sejnowski, T.J. (1983). Parallel visual computation. *Nature 306,* 21–26.

Barlow, H.B. (1981). Critical limiting factors in the design of the eye and visual cortex. *Proc. Roy. Soc.* (London) *212,* 1–34.

Beaulieu, C., and Colonnier, M. (1985). A laminar analysis of the number of round-asymmetrical and flat-symmetrical synapses on spines, dendritic trunks, and cell bodies in area 17 of the cat. *J. Comp. Neurol. 231,* 180–189.

Bernardo, K.L., and Woolsey, T.A. (1987). Axonal trajectories between mouse somatosensory thalamus and cortex. *J. Comp. Neurol. 258,* 542–564.

Blasdel, G.G., Lund, J.S., and Fitzpatrick, D. (1985). Intrinsic connections of macaque striate cortex: Axonal projections of cells outside lamina 4C. *J. Neurosci. 5,* 3350–3369.

Bonhoeffer, T. (1988). Synaptische Plastizität in Hippocampus-Slice-Kulturen. Untersuchungen mit Hilfe von 'Optical Recording' und Spannungsempfindlichen Flueoreszenzefarbstoffen. Thesis, University of Tübingen, FDR.

Braitenberg, V. (1978). Cortical architectonics: General and areal. In: *Architectonics of the Cerebral Cortex* (M.A.B. Brazier and H. Petsche, eds.). Raven Press, New York, pp. 443–465.

Brodmann, K. (1909). Vergleichende Lokalisationslehre der Grosshirnrinde in Ihren Prinzipien dargestellt auf Grund des Zellenbaues, Barth, Leipzig. Cited in Kemper and Galaburda (1984). Principles of cytoarchitectonics. In: *The Cerebral Cortex,* Vol. 1. Plenum Press, New York, pp. 35–57.

Burke, R.E. (1981). Motor units: Anatomy, physiology, and functional organization. In *Handbook of Physiology.* Vol. II, Motor Control, Part 1 (S.R. Geiger, ed.). American Physiological Society, Bethesda, Md., pp. 345–422.

Carney, T., Shadlen, M., and Switkes, E. (1987). Parallel processing of motion and colour information. *Nature 328,* 647–649.

Chapin, J.K. (1986). Laminar differences in sizes, shapes, and response profiles of cutaneous receptive fields in the rat SI cortex. *Exp. Brain Res. 62,* 549–559.

Coleman, P.D., and Riesen, A.H. (1968). Environmental effects on cortical dendritic fields. I. Rearing in the dark. *J. Anat.* (London) *102,* 363–374.

Crawford, M.L.J. (1985). Stimulus-specific columns in monkey visual cortex as revealed by the [^{14}C]-2-deoxyglucose method. In: *The Cerebral Cortex.* Vol. 3, Visual Cortex (A. Peters and E.G. Jones, eds.). Plenum Press, New York. pp. 331-349.

Crawford, M.L.J., Meharg, L.S., and Johnston, D.A. (1982). Structure of columns in monkey striate cortex induced by luminant-contrast and color-constrast stimulation. *Proc. Natl. Acad. Sci. USA 79,* 6722–6726.

Crick, F.H.C. (1984). Function of the thalamic reticular complex: The searchlight hypothesis. *Proc. Natl. Acad. Sci. USA 81,* 4586–4590.

DeFelipe, J., and Jones, E.G. (1985). Vertical organization of gamma-aminobutyric acid-accumulating intrinsic neuronal systems in monkey cerebral cortex. *J. Neurosci. 5,* 3246–3260.

Delacour, J., Houcine, O., and Talbi, B. (1987). "Learned" changes in the responses of the rat barrel field neurons. *Neurosci. 23,* 63–71.

Descarries, L., Watkins, K.C., and Lapierre, Y., (1977). Noradrenergic axon terminals in the cerebral corex of the rat. II. Topometric ultrastructural analysis. *Brain Res. 113,* 197–222.

DiPrisco, G.V. (1984). Hebb Synaptic Plasticity. *Prog. Neurol. 22,* 89–102.

Dow, B.M. (1978). Central mechanisms of vision: Parallel processing. *Fed. Proc. 35,* 54–59.

Dowling, J.E. (1968). Synaptic organization of the frog retina: An electron microscopic analysis comparing the retina of frogs and primates. *Proc. Roy. Soc.* ser. B. 170, 205–228.

Dowling, J.E., and Boycott, B.B. (1965). Neural connections in the retina: Fine structure of the IPL. *Cold Spring Harbor Symposia on Quantitative Biology 30,* 393–401.

Dowling, J.E., and Boycott, B.B. (1966). Organization of the primate retina: Electron microscopy. *Proc. Roy. Soc.* ser. B. 166, 80–111.

Dreher, B., Leventhal, A.G., and Hale, P.T. (1980). Geniculate input to cat visual cortex: A comparison of area 19 with areas 17 and 18. *J. Neurophysiol. 44,* 804–826.

Dykes, R.W. (1983). Parallel processing of somatosensory information: A theory. *Brain Res. Rev., 6,* 47–115.

Dykes, R.W., Landry, P., Metherate, R., and Hicks, T. P. (1984). Functional role of GABA in cat primary somatosensory cortex: Shaping receptive fields of cortical neurons. *J. Neurophysiol. 52,* 1066–1093.

Eccles, J.C. (1984). The cerebral neocortex: A theory of its operation. In: *Cerebral Cortex.* Vol. 2, Functional Properties of Cortical Cells (E.G. Jones and A. Peters, eds). Plenum Press, New York, pp. 1–36.

Edelman, G.M. (1978). Group selection and phasic reentrant signaling: A theory of higher brain function. In: *The Mindful Brain* (G.M. Edelman and V.B. Mountcastle, eds.). MIT Press, Cambridge, Mass., pp. 51–100.

Edelman, G.M. (1981). Group selection as the basis for higher brain function. In: *The Organization of the Cerebral Cortex* (G. Adelman, S.G. Dennis, F.O. Schmitt, and F.G. Worden, eds.), MIT Press, Cambridge, Mass., pp. 535–563.

Famiglietti, E.V. (1970). Dendrodendritic synapses in the lateral geniculate nucleus of the cat. *Brain Res. 20,* 181–191.

Famiglietti, E.V., Jr., and Peters, A. (1972). The synaptic glomerulus and the intrinsic neuron in the dorsal lateral geniculate nucleus of the cat. *J. Comp. Neurol. 144,* 285–334.

Foote, S.L., and Morrison, J.H. (1987). Extrathalamic modulation of cortical function. *Ann. Rev. Neurosci. 10,* 67–95.

Freund, T.F., Martin, K.A.C., Somogyi, P., and Whitebridge, D. (1985). Innervation of cat visual areas 17 and 18 by physiologically identified X- and Y-type afferents. II. Identification of postsynaptic targets by GABA immunocytochemistry and Golgi impregnation. *J. Comp. Neurol. 242,* 275–291.

Gabbott, P.L.A., Martin, K.A.C., and Whitteridge, D. (1987). Connections between pyramidal neurons in layer 5 of cat visual cortex (area 17). *J. Comp. Neurol. 259,* 364–381.

Garey, L.J., and Powell, T.P.S. (1971). An experimental study of the termination of the lateral geniculo-cortical pathway in the cat and monkey. *Proc. Roy. Soc.* ser. B. 179, 21–40.

Gilbert, C.D. (1977). Laminar differences in receptive field properties in cat primary visual cortex. *J. Physiol. 268,* 391–421.

Gilbert, C.D., and Wiesel, T.N. (1979). Morphology and intracortical projections of functionally characterised neurones in the cat visual cortex. *Nature* (London) *280,* 120–125.

Gilbert, C.D., and Wiesel, T.N. (1981). Laminar specialization an intracortical connections in cat primary visual cortex. In: *The Organization of the Cerebral Cortex* (G. Adelman, S.G. Dennis, F.O. Schmidt, F.G. Worden, eds.). MIT Press, Cambridge, Mass., pp. 163–191.

Globus, A., and Scheibel, A.B. (1967). The effect of visual deprivation on cortical neurons: A Golgi study. *Exp. Neurol. 19,* 331–345.

Gustafsson, B., Wingström, H., Abraham, W.C., and Huang, Y.-Y. (1987). Long-term potentiation in the hippocampus using depolarized current pulses as the conditioning stimulus to single volley synaptic potentials. *J. Neurosci. 7,* 774–780.

Hammond, P., and MacKay, D.M. (1981). Modulatory influences of moving textured backgrounds on responsiveness of simple cells in feline striate cortex. *J. Physiol.* (London) *319,* 431–442.

Hamos, J.E., Van Horn, S.C., Raczkowski, D., and Sherman, S.M. (1987). Synaptic circuits involving an individual retinogeniculate axon in the cat. *J. Comp. Neurol. 259,* 165–192.

Harvey, A.R. (1980). A physiological analysis of subcortical and commissural projections of areas 17 and 18 of the cat. *J. Physiol. 302,* 507–534.

Hebb, D.O. (1949). *The Organization of Behavior: A Neurophysiological Theory.* Wiley, New York.

Henry, G.H. (1977). Receptive field classes of cells in the striate cortex of the cat. *Brain Res. 133,* 1–28.

Hicks, T.P., and Dykes, R.W. (1983). Receptive field size of certain neurons in primary somatosensory cortex is determined by GABA-mediated intracortical inhibition. *Brain Res. 274,* 160–164.

Hoffman, K.P., and Stone, J. (1971). Conduction velocity of afferents to cat visual cortex: A correlation with cortical receptive field properties. *Brain Res. 32,* 460–466.

Hubel, D.H., and Livingstone, M.S. (1987). Segregation of form, color, and stereopsis in primate area 18. *J. Neurosci. 7,* 3378–3415.

Hubel, D.H., and Wiesel, T.N. (1962). Receptive fields, binocular interaction and functional architecture in the cat's visual cortex. *J. Physiol* (London) *160,* 106–154.

Hubel, D.H., and Wiesel, T.N. (1965). Receptive fields and functional architecture in two nonstriate visual areas (18 and 19) of the cat. *J. Neurophysiol. 28,* 229–289.

Hubel, D.H., and Wiesel, T.N. (1968). Receptive fields and functional architecture of monkey striate cortex. *J. Physiol. 195,* 215–243.

Hubel, D.H., and Wiesel, T.N. (1972). Laminar and columnar distribution of geniculocortical fibers in the macaque monkey. *J. Comp. Neurol. 146,* 421–450.

Hubel, D.H., and Wiesel, T.N. (1977). Functional architecture of macaque monkey visual cortex. *Proc. Roy. Soc.* ser. B. 198, 1–59.

Humphrey, A.L., Sur, M., Uhlrich, D.J., and Sherman, S.M. (1985). Projection patterns of individual X- and Y-cell axons from the lateral geniculate nucleus to cortical area 17 in the cat. *J. Comp. Neurol. 233,* 159–189.

Ito, M. (1985). Processing of vibrissa sensory information within rat neocortex. *J. Neurophysiol. 54,* 479–490.

Jones, E.G., and Powell, T.P.S. (1969). The cortical projection of the ventroposterior nucleus of the thalamus in the cat. *Brain Res. 13,* 298–318.

Juliano, S.L., and Whitsel, B.L. (1987). A combined 2-deoxyglucose and neurophysiological study of primate somatosensory cortex. *J. Comp. Neurol. 263,* 514–525.

Kandel, E.R. (1985). Processing of form and movement in the visual system. In: *Principles of Neuroscience* (E.R. Kandel and J.H. Schwartz, eds.). Elsevier, New York, pp. 366–383.

Kandel, E.R., Klein, M., Hochner, B., Shuster, M., Siegelbaum, S.A., Hawkins, R.D., Glandmanz, D.L., Catellucci, V.F., and Abrams, T.W.

(1987). Synaptic modulation and learning: New insights into synaptic transmission from the study of behavior. In: *Synaptic Function* (G.M. Edelman, W.E. Gall, W.M. Cowan, eds.). Wiley, New York, pp. 471–518.

Keller, A., and White, E.L. (1987). Synaptic organization of GABAergic neurons in the mouse SmI cortex. *J. Comp. Neurol. 262,* 1–12.

Kelso, S.R., Ganong, A.H., and Brown, T.H. (1986). Hebbian synapses in hippocampus. *Proc. Natl. Acad. Sci. USA 83,* 5326–5330.

Kennedy, C., Des Rosiers, H.M., Sakurada, O., Shinohara, M., Reivich, M., Jehle, J.W., and Sokoloff, L. (1976). Metabolic mapping of the primary visual system of the monkey by means of the autoradiographic [^{14}C]deoxyglucose technique. *Proc. Natl. Acad. Sci. USA 73,* 4230–4234.

Kidd, M. (1962). Electron microscopy of the internal plexiform layer of the retina in the cat and pidgeon. *J. Anat.* (London) *96,* 179–187.

Killackey, H.P., Belford, G., Ryugo, F., and Ryugo, D.K. (1976). Anomalous organization of thalamocortical projections consequent to vibrissae removal in the newborn rat and mouse. *Brain Res. 104,* 309–315.

Killackey, H.P., Ivy, G.O., and Cunningham, T.J. (1978). Anomalous organization of SMI somatotopic map consequent to vibrissae removal in the newborn rat. *Brain Res. 155,* 136–140.

Lennie, P. (1980). Parallel visual pathways: A review. *Vision Res. 20,* 561–594.

LeVay, S. (1986). Synaptic organization of claustral and geniculate afferents to the visual cortex of the cat. *J. Neurosci. 6,* 3564–3575.

Linsker, R. (1986a). From basic network principles to neural architecture: Emergence of orientation-selective cells. *Proc. Natl. Acad. Sci. USA 83,* 8390–8394.

Linsker, R. (1986b). From basic network principles to neural architecture: Emergence of orientation columns. *Proc. Natl. Acad. Sci. USA 83,* 8779–8783.

Livingstone, M.S., and Hubel, D.H. (1987). Psychophysical evidence for separate channels for the perception of form, color, movement, and depth. *J. Neurosci. 7,* 3416–3468.

Lorente de Nó, R. (1922). La corteza cerebral del raton. *Trab. Lab. Invest. Biol. Univ. Madrid 20,* 1–38.

Lorente de Nó, R. (1938). The physiology of the nervous system. In: *The Physiology of the Nervous System* (J.F. Fulton, ed.). pp. 291–329. Oxford Univ. Press, New York.

Lund, R.D. (1969) Synaptic patterns of the superficial layers of the superior colliculus of the rat. *J. Comp. Neurol.* 135, 179–208.

Lund, J.S., Hendrickson, A.E., Ogren, M.P., and Tobin, E.A. (1981). Anatomical organization of primate visual cortex area VII. *J. Comp. Neurol. 202,* 19–45.

Lund, J.S., Henry, G.H., MacQueen, C.L., and Harvey, A.R. (1979). Anatomical organization of the primary visual cortex (area 17) of the cat. A comparison with area 17 of the Macaque monkey. *J. Comp. Neurol. 184,* 599–618.

Magleby, K.L. (1987). Short-term changes in synaptic efficacy. In: *Synaptic Function* (G.M. Edelman, W.E. Gall, and W.M. Cowan, eds.). Wiley, New York, pp. 21–56.

Marrocco, R.T., Lane, R.F., McClurkin, J.W., Blaha, C.D., and Alkire, M.F. (1987). Release of cortical catecholamines by visual stimulation requires activity in thalamocortical afferents of monkey and cat. *J. Neurosci. 9,* 2756–2767.

Martin, K.A.C. (1984). Neuronal circuits in cat striate cortex. In: *The Cerebral Cortex.* Vol. II (E.G. Jones and A. Peters eds.) Plenum Press, New York, pp. 278–284.

McCasland, J.S., and Woolsey, T.A. (1988). High-resolution 2-Deoxyglucose mapping of functional cortical columns in mouse barrel cortex. *J. Comp. Neurol. 278,* 555–569.

McGuire, B.A., Hornung, J.-P., Gilbert, C.D., and Wiesel, T.N. (1984). Patterns of synaptic input to layer 4 of cat striate cortex. *J. Neurosci. 4,* 3021–3033.

McMullen, N.T., Goldberger, B., Suter, C.M., and Glaser, E.M. (1988). Neonatal deafening alters nonpyramidal dendrite orientation in auditory cortex: A computer microscope study in the rabbit. *J. Comp. Neurol. 267,* 92–106.

Merzenich, M.M., and Kaas, J.H. (1982). Reorganization of mammalian somatosensory cortex following preipheral nerve injury. *Trends in Neurosci.,* December, 434–436.

Merzenich, M.M., Kass, J.H., Wall, J.T., Sur, M., Nelson, R.J., and Felleman, D.J. (1983). Progression of change following median nerve section in the cortical representation of the hand in areas 3b and 1 in adult owl and squirrel monkeys. *Neuroscience 10,* 639–665.

Merzenich, M.M., Nelson, R.J., Stryker, M.P., Cynader, M.S., Schoppmann, A., and Zook, J.M. (1984). Somatosensory cortical map changes following digit amputation in adult monkeys. *J. Comp. Neurol. 224,* 591–605.

Michael, C.R. (1978a). Color vision mechanisms in monkey striate cortex: Simple cells with dual opponent-color receptive fields. *J. Neurophysiol. 41,* 1233–1249.

Michael, C.R. (1987b). Color-sensitive complex cells in monkey striate cortex: Simple cells with dual opponent-color receptive fields. *J. Neurophysiol. 41,* 1250–1266.

Morest, D.K. (1971). Dendrodendritic synapses of cells that have axons: The fine structure of the golgi type II cell in the medial geniculate body of the cat. *Z. Anat. Entwickl.-Gesch. 133,* 216–246.

Palm, G., and Bonhoeffer, T. (1984). Parallel processing for associative and neuronal networks. *Biol. Cybern. 51,* 201–204.

Parnevelas, J.G. (1984). Physiological properties of identified neurons. In: *Cerebral Cortex.* Vol. 2, Functional Properties of Cortical Cells (E.G. Jones and A. Peters, eds.), Plenum Press, New York, pp. 205–239.

Pearson, J.C., Finkel, L.H., and Edelman, G.M. (1987). Plasticity in the organization of adult cerebral cortical maps: A computer simulation based on neuronal group selection. *J. Neurosci. 7,* 4209–4223.

Peters, A. (1979). Thalamic input to the cerebral cortex. *T.I.N.S.* July, 183–185.

Peters, A. (1987) Number of neurons and synapses in primary visual cortex. In: Cerebral Cortex. Vol 6: Further Aspects of Cortical Function, Including

Hippocampus (eds. E.G. Jones and A. Peters), Plenum Press, New York, pp. 267–294.

Peters, A., Kara, D.A., and Harriman, K.M. (1985). The neuronal composition of area 17 of rat visual cortex. III. Numerical considerations. *J. Comp. Neurol. 238,* 263–274.

Pinching, A.J., and Powell, T.P.S. (1971). The neuropil of the glomeruli of the olfactory bulb. *J. Cell. Sci. 9,* 347–377.

Pons, T.P., Garraghty, P.E., Friedman, D.P., and Mishkin, M. (1987). Physiological evidence for serial processing in somatosensory cortex. *Science 237,* 417–419.

Ralston, H.J., and Herman, M.M. (1969). The fine structure of neurons and synapses in the ventrobasal thalamus of the cat. *Brain Res. 14,* 77–98.

Raviola, G., and Raviola, E. (1970). Light and electron microscopic observations on the internal plexiform layer of the rabbit retina. *Am. J. Anat. 120,* 403–426.

Reese, T.S., and Brightman, M.W. (1970). Olfactory surface and central olfactory connections in some vertebrates. In: *C.I.B.A. Foundation, Taste and Smell in Vertebrates.* (G.E.W. Wolstenholme and J. Knight, eds.). J. and A. Churchill, London, pp. 115–149.

Rosenquist, A.C. (1985). Connections of visual cortical areas in the cat. In: *The Cerebral Cortex.* Vol. 3, Visual Cortex (A. Peters and E.G. Jones, eds.). Plenum Press, New York, pp. 81–117.

Sejnowski, T.J. (1986). Open questions about computation in cerebral cortex. In: *Parallel Distributed Processing: Explorations in the Microstructure of Cognition.* Vol. 2, Psychological and Biological Models (J.L. McClelland, D.E. Rumelhart, and the PDP Research Group, eds.). MIT Press, Cambridge, Mass., pp. 372–389.

Shepherd, G.M. (1979). *The Synaptic Organization of the Brain.* 2nd ed. Oxford Univ. Press, New York.

Sholl, D.A. (1956). *The Organization of the Cerebral Cortex* (Facsimile of the 1956 Edition). Hafner Publishing Co., London, 1967.

Sillito, A.M. (1975). The contribution of inhibitory mechanisms to the receptive field properties of neurones in the striate cortex of the cat. *J. Physiol.* (London) *250,* 305–329.

Sillito, A.M. (1977). Inhibitory processes underlying the directional specificity of simple, complex and hypercomplex cells in the cat's visual cortex. *J. Physiol.* (London) *271,* 699–720.

Sillito, A.M. (1984). Functional considerations of the operation of GABAergic inhibitory processes in the visual cortex. In: *The Cerebral Cortex.* Vol. 2, Functional Properties of Cortical Cells (E.G. Jones and A. Peters, eds.). Plenum Press, New York, pp. 91–117.

Silverman, A.-J., Oldfield, B., Hou-Yu, A., and Zimmerman, E.A. (1985). The noradrenergic innervation of vasopressin neurons in the paraventricular nucleus of the hypothalamus: An ultrastructural study using radioautography and immunocytochemistry. *Brain Res. 325,* 215–229.

Simons, D.J. (1978). Response properties of vibrissa units in rat SI somatosensory neocortex. *J. Neurophysiol. 41,* 798–820.

Simons, D.J. (1985). Temporal and spatial integration in the rat SI vibrissae cortex. *J. Neurophysiol. 54,* 615–635.

Simons, D.J., and Woolsey, T.A. (1979). Functional organization in mouse barrel cortex. *Brain Res. 165,* 327–332.

Singer, W., Tretter, F., and Cynader, M. (1975). Organization of cat striate cortex: A correlation of receptive-field properties with afferent and efferent connections. *J. Neurophysiol. 38,* 1080–1098.

Sloper, J.J. (1973). An electron microscope study of the termination of afferent connections to the primate motor cortex. *J. Neurocytol. 2,* 361–368.

Sloper, J.J., and Powell, T.P.S. (1979). An experimental electron microscopic study of afferent connections to the primate motor and somatic sensory cortices. *Phil. Trans. Roy. Soc.* ser. B. 285, 199–225.

Somogyi, P., Cowey, A., Halasz, N., and Freund, T.F. (1981). Vertical organization of neurons accumulating 3H-GABA in the visual cortex of the rhesus monkey. *Neurosci.* 7, 2577–2609.

Somogyi, P., Cowey, A., Kisvarday, Z.F., Freund, T.F., and Szentagothai, J. (1983). Retrograde transport of γ-amino[^{3}H]butyric acid reveals specific interlaminar connections in the striate cortex of monkey. *Proc. Natl. Acad. Sci. USA 80,* 2385–2389.

Stone, J., Dreher, B., and Leventhal, A. (1979). Hierarchical and parallel mechanisms in the organization of the visual cortex. *Brain Res. Rev. 1,* 345–394.

Suga, N. (1984). The extent to which biosonar information is represented in the bat auditory cortex. In: *Dynamic Aspects of Neo-Cortical Function* (G.M. Edelman, W.E. Gall, and W.M. Cowan, eds.). Wiley, New York, pp. 315–373.

Suga, N. (1987). Echolocation. In: *Encyclopedia of Neuroscience* (G. Adelman, ed.). Birkhauser, Boston, Mass., pp. 360–364.

Suga, N., and Horikawa, J. (1986). Multiple time axes for representation of echo delays in the auditory cortex of the mustached bat. *J. Neurophysiol. 55,* 776–805.

Symonds, L.L., and Rosenquist, A.C. (1984). Laminar origins of visual corticocortical connections in the cat. *J. Comp. Neurol. 229,* 39–47.

Szentágothai, J. (1979). Local neuron circuits of the neocortex, In: *The Neurosciences Fourth Study Program* (F.O. Schmitt and F.G. Worden, eds.). MIT Press, Cambridge, Mass., pp. 39–45.

Toyama, K., Maekawa, K., and Takeda, T. (1977). Convergence of retinal inputs onto visual cortical cells: I. A study of the cells monosynaptically excited from the lateral geniculate body. *Brain Res., 137,* 207–220.

Vaadia, E., Gotlieb, Y., and Abeles, M. (1982). Single unit activity related to sensorimotor association in auditory cortex of monkey. *J. Neurophysiol. 48,* 1201–1213.

Valverde, F. (1968). Structural changes in the area striata of the mouse after enucleation. *Exp. Brain Res. 5,* 274–292.

Van Essen, D.C. (1985). Functional organization of primate visual cortex. In: *Cerebral Cortex.* Vol. 3, Visual Cortex (Peters, A. and Jones, E.G., eds.). Plenum Press, New York, pp. 259–329.

Van Essen, D.C., and Maunsell, J.H.R. (1980). Two-dimensional maps of the cerebral cortex. *J. Comp. Neurol. 191,* 255–281.

White, E.L. (1972). Synaptic organization in the olfactory glomerulus of the mouse. *Brain Res. 37,* 69–80.

White, E.L. (1973). Synaptic organization of the mammalian olfactory glomerulus: New findings including an intraspecific variation. *Brain Res. 40,* 299–313.

White, E.L. (1978). Identified neurons in mouse SmI cortex which are postsynaptic to thalamocortical axon terminals: A combined Golgi-electron microscopic and degeneration study. *J. Comp. Neurol. 181,* 627–662.

White, E.L. (1979). Thalamocortical synaptic relations: A review with emphasis on the projection of specific thalamic nuclei to the primary sensory areas of the neocortex. *Brain Res. Rev. 1,* 275–311.

White, E.L. (1981). Thalamocortical synaptic relations. In: *The Organization of the Cerebral Cortex* (G. Adelman, S.G. Dennis, F.O. Schmitt, and F.G. Worden, eds.). MIT Press, Cambridge, Mass., pp. 153–161.

White, E.L. (1986). Terminations of thalamic afferents. In: *The Cerebral Cortex.* Vol. 5, Sensory-Motor Areas and Aspects of Cortical Connectivity (E.G. Jones and A. Peters, eds.) Plenum Press, New York, pp. 271–289.

White, E.L. (1987). Thalamocortical interactions. In: *Encyclopedia of Neuroscience.* Vol. II (George Adelman, ed.) Birkhaüser Press, Boston, Mass., pp. 1202–1204.

White, E.L., Benshalom, G., and Hersch, S.M. (1984). Thalamocortical and other synapses of non-spiny multipolar cells in mouse SmI cortex. *J. Comp. Neurol. 229,* 311–320.

White, E.L., and Hersch, S.M. (1982). A quantitative study of thalamocortical and other synapses involving the apical dendrites of corticothalamic projection cells in mouse SmI cortex. *J. Neurocytol. 11,* 137–157.

White, E.L. and Keller, A. (1987). Intrinsic circuitry involving the local axonal collaterals of corticothalamic projection cells in mouse SmI cortex. *J. Comp. Neurol. 262,* 13–26.

Winfield, D.A., and Powell, T.P.S. (1983). Laminar cell counts and geniculocortical boutons in area 17 of cat and monkey. *Brain Res. 277,* 223–229.

Woolsey, T.A., Durham, D., Harris, R.M., Simons, D.J., and Valentino, K.L. (1981). Somatosensory development. In: *Development of Perception,* Vol. 1. Academic Press, New York, pp. 259–292.

Zarzecki, P. (1985). Convergent inputs to identified sonsory cortex neurons. In: *Development, Organization and Processing in Somatosensory Pathways.* Alan R. Liss, New York, pp. 259–263.

Index